GENDER EQUALITY AND SOCIAL INCLUSION IN PUBLIC WATER, SANITATION, AND HYGIENE FACILITIES IN DEVELOPING ASIA AND THE PACIFIC

NOVEMBER 2024

ASIAN DEVELOPMENT BANK

CONTENTS

CHECKLISTS, TABLES, FIGURES, AND BOXES

Checklists

Tables

Figures

Boxes

ABBREVIATIONS

ADB	Asian Development Bank
GDP	gross domestic product
GESI	gender equality and social inclusion
HRBA	human rights-based approach
JMP	Joint Monitoring Programme for Water Supply, Sanitation and Hygiene
NBC	National Building Code
NDP	National Development Plan
O&M	operation and maintenance
SDG	Sustainable Development Goal
SDP	Sector Development Plan
SOGIESC	sexual orientation, gender identity, gender expression, and sex characteristics
UN	United Nations
UNICEF	United Nations Children's Fund
WASH	water, sanitation, and hygiene
WHO	World Health Organization

ACKNOWLEDGMENTS

This publication was prepared by Kate Walton (consultant), with guidance and insights from the Water and Urban Development Sector Office, Sectors Group (SG-WUD) of the Asian Development Bank (ADB), under the leadership of Satoshi Ishii (Director, SG-WUD).

The team extends its gratitude to the following ADB colleagues for their invaluable input and review during the development of this publication: Neeta Pokhrel (Director, SG-WUD), Samantha Hung (Director, Climate Change and Sustainable Development Department [CCSD]), Alexandra Conroy (Senior Urban Development Specialist, SG-WUD), Jitendra Singh (Water Supply and Sanitation Specialist, SG-WUD), Shiva Prasad Paudel (Project Officer, SG-WUD), Lewelyn Baguyo (Social Development Officer, CCSD), Malika Shagazatova (Senior Social Development Specialist, CCSD), Rosemary Victoria Atabug (Senior Gender Officer, CCSD), Saswati Belliappa (Senior Safeguards Specialist, Office of Safeguards), Shuvechha Khadka (Senior Social Development Officer, Nepal Resident Mission), and Mafalda Pinto (former Water and Sanitation Specialist, SG-WUD).

The team also acknowledges the support provided by Fatima Bautista (Associate Water Resource Officer, SG-WUD), Graziel Salazar (consultant), Jason Beerman (consultant), and Marie Christelle Garcia (consultant) in coordinating and overseeing the publication.

Please note that the information in this publication is based on research and analysis and is not intended as professional or legal advice. It does not create new legal obligations or address specific impacts. While the analysis serves its intended purposes, it does not establish facts or liability, and not all contributors may endorse every part. Changes occurring after publication may affect the accuracy of the information.

This publication was developed through the technical assistance project "Accelerating Sanitation for All in Asia and the Pacific," supported by the Sanitation Financing Partnership Trust Fund.

EXECUTIVE SUMMARY

Since 2000, many developing countries in Asia and the Pacific have significantly improved access to water, sanitation, and hygiene (WASH). Access to at least basic WASH services has increased substantially, and many countries have made notable progress toward achieving the Sustainable Development Goals related to WASH.

Efforts have largely focused on improving household-level WASH access. While crucial, this has sometimes resulted in the neglect of WASH services in public spaces. As a result, schools and health care facilities often lack adequate WASH infrastructure, and key public spaces such as transport hubs, marketplaces, parks, and government buildings frequently fail to meet users' sanitation and hygiene needs.

The lack of accessible and adequate public WASH facilities limits people's ability to participate in essential activities. Without access to safely managed toilets, water, and hygiene facilities, individuals lose work and income-generating opportunities, miss school, or forego seeking health care.

Although poor WASH access affects everyone, certain groups face greater challenges in managing basic bodily functions when facilities are inaccessible or inadequate, especially in public spaces. These impacts, though difficult to quantify, are no less real. They stem from the failure to recognize and address the diverse WASH needs of different social groups in the design, improvement, and maintenance of facilities.

While specific needs vary, particular attention should be given to the following groups when providing public WASH facilities:

(i)   children;
(ii)   women, girls, and other people who menstruate;
(iii)   people with disabilities;
(iv)   people with diverse sexual orientation, gender identity, gender expression, and sex characteristics;
(v)   older people;
(vi)   carers and caregivers, including parents and guardians; and
(vii)   people experiencing poverty.

Thus, it is crucial that public WASH facilities incorporate gender equality and social inclusion (GESI) principles in their design and operation and maintenance. For example, women, girls, and other people who menstruate need water for personal cleaning and hygienic disposal options for menstrual products. People with disabilities require facilities with features such as ramps, larger cubicles, handrails, and tactile paving. Carers of babies and young children need spaces for changing diapers and safe disposal options.

Standards for public WASH facilities are generally outlined in national building codes. Many nations also have stand-alone WASH regulations, policies, and standards, along with sector development plans that cover short-, medium-, and long-term goals. Some countries have developed specific guidelines or standards for schools and health care facilities, used in conjunction with national building codes, to establish minimum service levels for WASH infrastructure.

National WASH guidelines or standards serve as the primary reference for designing or upgrading public WASH facilities. However, as our understanding of GESI evolves, existing documents may not cover all components. This publication provides guidance on integrating GESI components while recognizing the diversity of local contexts and the complexity of achieving universal WASH coverage.

At a minimum, public WASH facilities should

(i)   provide gender-segregated and gender-neutral facilities;
(ii)   ensure safety and privacy;
(iii)   provide water for toileting, washing, and handwashing;
(iv)   incorporate disability-friendly and universal design features;
(v)   support the management of menstruation; and
(vi)   accommodate the needs of parents and caregivers.

In addition, public WASH facilities must be supported by sufficient funding, human resources, and robust operation and maintenance procedures to ensure that adequate WASH facilities are accessible to all in public spaces.

INTRODUCTION   1

In recent years, developing countries in Asia and the Pacific have made significant strides toward improving access to quality water, sanitation, and hygiene (WASH) facilities in both households and public spaces. Despite this substantial progress, challenges persist, particularly in public spaces, where WASH facilities often remain inadequate and fail to meet users' needs, especially those of women and socially marginalized groups.

This publication provides recommendations for ensuring that public WASH facilities in developing Asia and the Pacific are not only gender-inclusive but also socially inclusive. It is intended for use by all stakeholders in water and sanitation, urban and rural development, education, health care, social services, and other public sectors. Additionally, it will be useful for the Asian Development Bank (ADB) staff and consultants in designing, implementing, and evaluating WASH programs.

This publication outlines ways to ensure that gender equality and social inclusion (GESI) is central to designing new public WASH facilities and improving existing ones. It features two case studies (Nepal and Fiji) and provides information and guidance on the

(i)   principles of GESI;
(ii)   impact of gender and social identity on access to WASH;
(iii)   key components of gender- and socially inclusive public WASH facilities, with particular attention to schools and health care facilities;
(iv)   operation and maintenance (O&M) of public WASH facilities; and
(v)   the importance of community engagement in the design and O&M.

Specifically, the recommendations and checklists in this publication are designed to assist in the design, planning, construction, renovation, O&M, and monitoring and evaluation of public WASH facilities to ensure they meet the needs of women, girls, people with disabilities, sexual and gender minorities, and other marginalized groups. The recommendations and checklists can be used to assess and improve existing public WASH facilities as well as to guide the development and construction of new ones.

This publication is not a one-size-fits-all set of guidelines, nor is it exhaustive. It draws from international and regional standards and research to outline GESI principles and provide recommendations on ensuring all people have access to public WASH facilities. It should be used alongside existing national guidelines, policies, and procedures to advance GESI goals. Additional resources are provided in Appendix 2.

ADB recognizes that working toward GESI is a complex process requiring significant time and resources, as well as attention to multiple levels and aspects of inclusion and intersectionality. As such, ADB urges all stakeholders to strive for more than the bare minimum and pursue the highest levels of inclusion possible.

2 GENDER AND SOCIAL INCLUSION IN WATER, SANITATION, AND HYGIENE

Access to safe and sustainable water and sanitation services is a human right and essential for all people in all aspects of life. Good WASH services improve health and education outcomes, expand economic participation, and enhance GESI. According to 2019 World Health Organization (WHO) estimates, improving access to WASH could save 1.4 million lives per year worldwide, and universal coverage could bring significant economic benefits.[1] Simply ensuring that everyone has access to a safely managed toilet could generate $86 billion a year in greater productivity and reduced health costs, as well as increase school and work attendance by 3 billion days annually.[2]

Despite the centrality of WASH to all elements of life, its importance is often neglected in public buildings and spaces. In many countries, the provision of WASH services continues to focus on private homes. Inadequate WASH facilities in schools, health care facilities, transport hubs, marketplaces, and government buildings can limit access to essential services and participation in daily activities. When these and other public buildings lack appropriate and sufficient WASH facilities, people often choose to avoid visiting or using their services. In practical terms, this means that people may forgo seeking health care, children may miss school, and income opportunities may be lost.

For women, girls, people with disabilities, older people, people with diverse sexualities and gender identities, and other marginalized or vulnerable groups, inadequate WASH facilities in public buildings have an even greater impact. Girls and children with disabilities may miss out on education when WASH facilities are unavailable or inadequate, especially during menstruation. Furthermore, unsafe WASH facilities in public buildings can expose users to sexual- and gender-based violence, including sexual exploitation, abuse, and harassment in workplaces. These barriers and challenges can be addressed by adopting a gender- and socially inclusive approach to WASH (Box 1).

---

[1]   WHO. 2023. *Burden of Disease Attributable to Unsafe Drinking-Water, Sanitation and Hygiene: 2019 Update.*
[2]   WaterAid. 2021. *Mission-Critical: Invest in Water, Sanitation and Hygiene for a Healthy and Green Economic Recovery.*

Box 1: What Is Gender- and Socially Inclusive Water, Sanitation, and Hygiene?

Gender- and socially inclusive water, sanitation, and hygiene (WASH) refers to services and facilities that address the needs of women, girls, people with disabilities, people with diverse sexualities and gender identities, and other marginalized or vulnerable groups. This includes individuals who face exclusion or discrimination due to ethnicity, indigeneity, caste, class, religion, health status, marital status, age, or other aspects of identity.

WASH is gender- and socially inclusive when the physical, biological, and social needs of these groups are actively considered and integrated into WASH service provision, as well as the design and operation and maintenance of WASH infrastructure.

In the context of public WASH facilities, gender- and socially inclusive WASH refers to facilities that are accessible to all and that enable everyone to manage their sanitation and hygiene needs safely, comfortably, and with dignity.

Source: Asian Development Bank.

Key Components of Water, Sanitation, and Hygiene

While water, sanitation, and hygiene components are intrinsically linked, each involves distinct elements. WASH falls under Sustainable Development Goal (SDG) 6, which aims to ensure the availability and sustainable management of water and sanitation for all by 2030. SDG 6 emphasizes addressing the needs of women, girls, and people in vulnerable situations.

Water

SDG 6.1 targets universal and equitable access to safe and affordable drinking water for all. This is measured using the indicator of "safely managed drinking water" developed by the WHO/United Nations Children's Fund (UNICEF) Joint Monitoring Programme for Water Supply, Sanitation and Hygiene (JMP) (Figure 1). Safely managed drinking water covers accessibility, availability, and quality. A safely managed drinking water service is (i) accessible on premises, (ii) available when needed, and (iii) free from fecal and chemical contamination.

Figure 1: Drinking Water Ladder

Surface Water	Unimproved	Limited	Basic	Safely Managed
Drinking water directly from a river, dam, lake, pond, stream, canal, or irrigation canal	Drinking water from an unprotected dug well or unprotected spring	Drinking water from an improved source for which collection time exceeds 30 minutes for a round trip (including queuing)	Drinking water from an improved source, provided collection time is not more than 30 minutes for a round trip (including queuing)	Drinking water from an improved water source that is accessible on premises, available when needed, and free from fecal and priority chemical contamination

Source: World Health Organization/United Nations Children's Fund. Joint Monitoring Programme: Drinking Water. https://washdata.org/monitoring/drinking-water.

Sanitation

Sanitation is covered by SDG 6.2, which targets access to adequate and equitable sanitation and hygiene for all while aiming to end open defecation. This goal includes facilities and services such as toilets and handwashing stations, which are crucial for improving health outcomes, especially for children. The JMP defines safely managed sanitation services as those provided by improved sanitation facilities that are not shared with other households and where waste is either treated and disposed of on-site, stored temporarily before being transported for off-site treatment, or transported through a sewer and treated off-site (Figure 2).

Figure 2: Sanitation Ladder

Open Defecation	Unimproved	Limited	Basic	Safely Managed
Disposal of human waste in fields, forests, bushes, open bodies of water, beaches, and other open spaces or with solid waste	Use of pit latrines without a slab or platform, hanging latrines, or bucket laterines	Use of improved facilities shared between two or more households	Use of improved facilities, which are not shared with other households	Use of improved facilities that are not shared with other households and where human waste is safely disposed of on-site or removed and treated off-site

Source: World Health Organization/United Nations Children's Fund. Joint Monitoring Programme: Sanitation. https://washdata.org/monitoring/sanitation.

Hygiene

Hygiene is also covered by SDG 6.2. It refers to the conditions and practices that help maintain health and prevent the spread of disease, including handwashing, food hygiene, and menstrual health and hygiene management. Among these, handwashing is the most monitored element of hygiene (Figure 3).

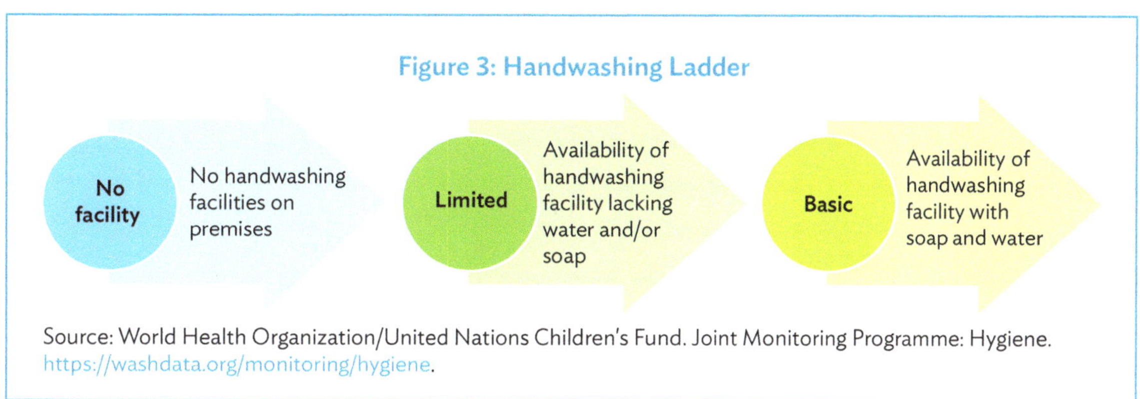

Figure 3: Handwashing Ladder

Source: World Health Organization/United Nations Children's Fund. Joint Monitoring Programme: Hygiene. https://washdata.org/monitoring/hygiene.

Menstrual health is key to good hygiene and overall health (Figure 4). While not specifically mentioned in the SDGs, it is closely linked to SDG 6.2, which aims to ensure access to adequate and equitable sanitation and hygiene for all, with particular attention to the needs of women and girls. Monitoring of menstrual health is improving, although as per the 2022 JMP report, only 53 countries collected data on menstrual health—up from 42 in 2020.

Figure 4: Good Menstrual Health

Source: Asian Development Bank. 2023. *Addressing Menstrual Health in Urban, Water, and Sanitation Interventions in the Pacific: Practitioner Guide.*

Gender Equality, Social Inclusion, and Leaving No One Behind

Gender equality refers to the equal rights, responsibilities, and opportunities of all genders. A gender equality approach requires that the interests, needs, and priorities of all genders be considered and that diversity be recognized among them. Gender equality efforts primarily target women and girls, aiming to shift the gender roles, norms, and expectations imposed on them. Some gender equality initiatives also focus on men and boys, depending on social situations and local contexts.

Social inclusion is the process of improving the ability, opportunity, and dignity of individuals to participate meaningfully in society. Social inclusion efforts focus on people who are marginalized, disadvantaged, vulnerable, or discriminated against based on their social identity. This may include factors such as gender, race, ethnicity, indigeneity, caste, class, religion, sexual orientation, disability, health status, marital status, or age.

GESI is central to sustainable development. It can be viewed as a continuum, from GESI-blind on the left to GESI-transformative on the right (Figure 5). The SDGs incorporate similar concepts, using terms like "universal" and "equitable." "Universal" means that everyone should have access to the necessary resources and services, including in public spaces and buildings. "Equitable" means the absence of discrimination, encapsulated by the explicit SDG principle to "leave no one behind."

GESI-blind (also called GESI-unaware or GESI-harmful) approaches. These approaches either do not recognize the different roles and needs of various groups within the community or recognize these differences but fail to address them. As a result, GESI-blind programs tend to perpetuate the current inequitable situation.

GESI-sensitive (also called GESI-aware) approaches. These approaches recognize that different groups within the community have distinct roles and needs and take some action to address this. However, inequalities are not specifically targeted, so GESI-sensitive approaches typically result in only minor changes, focusing on infrastructure and practical needs rather than on social structures or enabling environments.

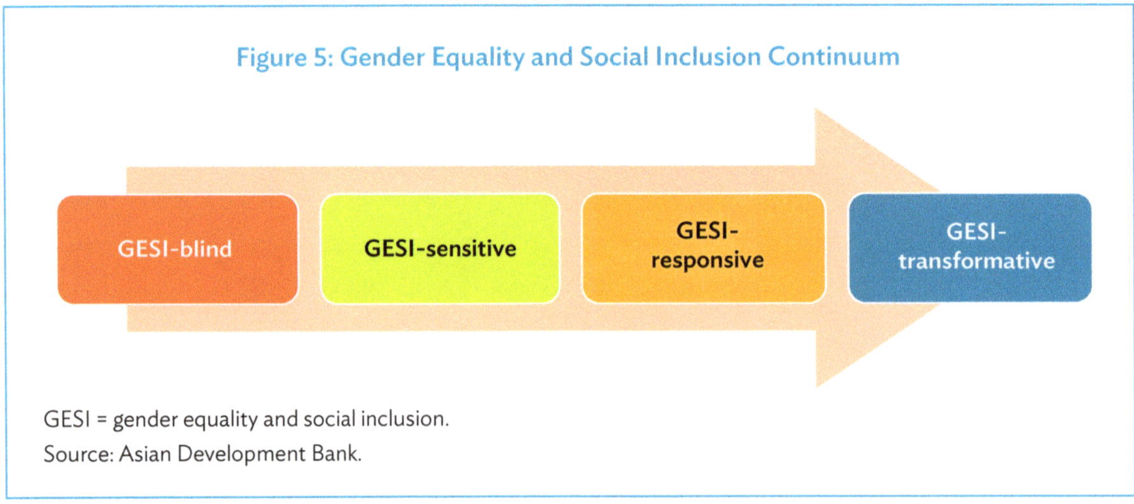

GESI = gender equality and social inclusion.
Source: Asian Development Bank.

GESI-responsive approaches. These approaches also recognize the varying needs of different groups within the community but go further by acknowledging the importance of addressing inequality, as well as the barriers to participation and representation of marginalized and vulnerable groups. GESI-responsive approaches address both practical needs and enabling environments (including social norms) but may lead to only surface-level or temporary results as deeper social change is often limited.

GESI-transformative approaches. These approaches take a further step by explicitly recognizing and challenging harmful social structures and norms to rectify power imbalances and improve the position of marginalized and vulnerable groups. Crucially, GESI-transformative approaches anticipate that pushing for social change can provoke resistance, so they include strategies to respond to backlash as part of the approach.

The key difference between GESI-responsive and GESI-transformative approaches is that GESI-responsive approaches focus on ensuring everyone has access to WASH facilities and services when and where they need them. In contrast, GESI-transformative approaches do this while explicitly challenging harmful social norms and power imbalances to bring about meaningful social change. Both approaches are useful in different contexts.

For a more in-depth resource on GESI in WASH, Water for Women and the Sanitation Learning Hub have developed the Gender Equality and Social Inclusion Self-Assessment Tool.[3] This tool helps identify specific, context-appropriate approaches and strategies that support GESI-transformative WASH.

How Gender and Social Identity Affect Access to Water, Sanitation, and Hygiene

A lack of adequate WASH facilities negatively impacts health, education, and the economy. It can lead to illness and death due to diarrheal diseases, soil-transmitted diseases, vector-borne diseases, antimicrobial resistance, and stunting. It is estimated that more than 1.9 million deaths worldwide could have been prevented in 2016 with adequate WASH. Economically, poor sanitation alone costs more than 2% of gross domestic product (GDP) in East Asia and the Pacific and more than 4% of GDP in South Asia.[4]

While poor access to WASH affects everyone, social identities and inequalities mean some groups face greater difficulty in managing their normal bodily functions, particularly in public spaces, if WASH facilities are inaccessible or fail to meet their needs. For example, women's behavior in using public toilets is shaped by inequality (Box 2).[5]

---

[3]   Water for Women and the Sanitation Learning Hub. 2021. *Gender Equality and Social Inclusion Self-Assessment Tool: Facilitation Guide for WASH Project Managers, Researchers and Self-Assessment Facilitators—Working Towards Transformation in Inclusive Water, Sanitation and Hygiene.*

[4]   UNICEF and WHO. 2020. *State of the World's Sanitation: An Urgent Call to Transform Sanitation for Better Health, Environments, Economies and Societies.*

[5]   Women in Warangal City ultimately preferred women caretakers. However, they were open to having a male caretaker over not having one at all. Thus, it is recommended to investigate the age of the male caretaker that women would feel comfortable with. Y.M. Reddy, V.S. Chary, and R. Srividya. 2017. *Why Do Women in India Not Use Public Toilets? Patterns and Determinants of Usage by Women in Warangal City.* Paper prepared for the 40th WEDC International Conference, Loughborough, United Kingdom.

Box 2: Why Do Women Not Use Public Toilets?

Women may avoid using public toilets for various reasons, depending on the context. An evaluation in Warangal, Telangana, India, found that women avoided public toilets because of dirtiness, inappropriate locations, the presence of men near entrances, and male caretakers.

Women preferred public toilets with running water and soap; female or trained older male caretakers; women-only toilet blocks; safer and more private access (not directly off the main road); toilets in or near gas stations, bus stops, and railway stations; a choice of squat and raised Western-style toilets; better availability; and proper disposal for sanitary pads.

The city government used this evaluation to design and build new public toilets, including four women-only toilets.

Source: WaterAid, Water and Sanitation for the Urban Poor, and United Nations Children's Fund. 2018. *Female-Friendly Public and Community Toilets: A Guide for Planners and Decision Makers.*

It is also important to note that people's identities and the inequalities they experience overlap. This concept is called "intersectionality." Intersectionality recognizes that people's lives are shaped by their identities, relationships, and social factors, which intersect with one another and with social structures of power, creating different forms of privilege and oppression.

Kimberlé Crenshaw, who coined the term, summarizes intersectionality as "a way of thinking about identity and its relationship to power."[6] Understanding intersectionality, specifically in how gender and social inclusion affect access to WASH, helps ensure that WASH policies, programs, and services are more inclusive and uphold the principle of leaving no one behind.

Children

WASH is crucial for children's health. Good access to WASH facilities ensures that children can learn, play, and grow in healthy environments. Inadequate WASH is a leading cause of diarrheal disease, responsible for the deaths of half a million children every year—accounting for 9% of all child deaths globally.[7]

WASH for children is particularly important in schools, where they spend up to a third of their time. Poor access to adequate and clean toilets and handwashing facilities can reduce attendance, hinder educational achievement, and spread illness. A study in Colombia found that children with access to water and soap were three times more likely to wash their hands at crucial times, significantly less likely

---

[6]   UN Women and United Nations Partnership on the Rights of Persons with Disabilities. 2022. *Intersectionality Resource Guide and Toolkit: An Intersectional Approach to Leave No One Behind.*

[7]   UNICEF. UNICEF Data. Diarrhoea. https://data.unicef.org/topic/child-health/diarrhoeal-disease/ (accessed 21 August 2023).

to report gastrointestinal illness, and 20% less likely to be absent from school. A lack of safe drinking water also negatively impacts students, resulting in dehydration, urinary tract infections, and difficulty concentrating in class or participating in sports.[8]

Interestingly, children can be powerful agents of change within their homes. They often bring home good hygiene and sanitation behaviors learned at school, such as handwashing, spreading positive behavior to their households and communities.

Special attention should also be given to babies, young children, and out-of-school children. Children with disabilities and those from marginalized communities—especially girls—are more likely to be out of school. In many developing countries, informal workers often bring their babies and young children to workplaces, so public WASH facilities should also accommodate children's needs in spaces like marketplaces and public transport hubs.[9] This includes providing child-appropriate toilets and sinks as well as facilities for adults of all genders to change babies' nappies and clothes.

Women and Girls

Women and girls are especially affected by inadequate public WASH facilities. Biologically, they need more frequent and longer access to toilets due to menstruation and pregnancy. Socially, caregiving responsibilities, especially for children, increase their reliance on public toilets. Culturally, expectations around dignity and modesty often further restrict their access to WASH facilities.

Failing to meet these needs restricts women's movement; increases social isolation; and reduces participation in public life, income generation, and education. WHO reports that 11% more girls attend school when proper sanitation is available.[10] Adult women, often caregivers for children, people with disabilities, the ill, and older people, frequently face unsupportive facilities in public. For example, narrow toilets make it difficult for women to assist others, and the lack of changing tables makes it impossible to hygienically and comfortably change children's diapers.

Public WASH facilities often fail to support women and girls in managing their menstrual needs.[11] Menstrual health is defined as "a state of complete physical, mental and social well-being and not merely the absence of disease or infirmity, in relation to the menstrual cycle."[12] Supporting good menstrual health is essential for gender equality, yet this need is overlooked in many public WASH facilities in both developing and developed countries. Using menstrual products for too long can risk overflow (leading to embarrassment), infection, or toxic shock syndrome. At schools, girls may avoid attending due to inadequate menstrual hygiene facilities such as rubbish bins, water, or privacy for changing and disposing of menstrual materials.

---

8   C. Jasper, T-T. Le, and J. Bartram. 2012. Water and Sanitation in Schools: A Systematic Review of the Health and Educational Outcomes. *International Journal of Environmental Research and Public Health.* 9 (8). pp. 2272–2787.

9   Water for Women. 2023. *Learning Brief: WASH in Schools—Insights from Water for Women.*

10 WHO. 2002. *World Health Report 2002: Reducing Risks, Promoting Healthy Life.*

11 In addition to women and girls, other individuals also menstruate, such as transgender men, nonbinary people, and third-gender people. The menstrual management needs of all should be considered when designing and improving public WASH facilities.

12 J. Hennegan et al. 2021. Menstrual Health: A Definition for Policy, Practice, and Research. *Sexual Reproductive Health Matters.* 29 (1).

Poor public WASH facilities can negatively impact the health of women and girls. Older women and those who have given birth are more likely to experience incontinence than men—affecting 1 in 4 women over 35, compared to 1 in 10 men—resulting in a greater need for frequent toilet use.[13] Meanwhile, in health care facilities, poor sanitation and hygiene can lead to reproductive infection, miscarriage, and preterm birth, putting both women and babies at serious risk.

The risk and fear of violence for women and girls increase when public WASH facilities are in unsafe or insufficiently private locations. As one female respondent expressed in a survey on women's attitudes toward using toilets in public, at school, and at work, "Often, bathrooms are in areas that are very isolated or bathrooms are empty, creating situations that can leave women vulnerable to assault."[14] In Bihar, India, for example, nearly half of all reported rapes in 2012 were "sanitation-related" (footnote 12), while in South Africa, schoolgirls feared sexual attacks when using toilets far from school buildings.[15] As a result, women and girls often avoid public WASH facilities, miss school during menstruation, or avoid going out altogether.

Finally, in 7 out of 10 households globally, women and girls over age 15 are primarily responsible for water collection, taking valuable time away from work, education, and leisure.[16]

People with Disabilities

About 15% of the world's population has some form of disability, and more than 700 million people with disabilities live in Asia and the Pacific alone.[17] Despite this, many people with disabilities face significant barriers in accessing WASH facilities, especially in public, as their specific needs are often overlooked during design and construction.

Barriers include lack of appropriate physical access; facilities that are difficult, dangerous, or impossible to use or too small for a carer to assist; lack of appropriate hygiene information; and social stigma, discrimination, harassment, and violence.[18]

As a result, people with disabilities are often excluded from public life, impacting not only individuals but also macro socioeconomic components such as GDP. The cost of excluding people with disabilities is estimated to represent up to 7% of GDP in some countries.[19]

---

[13]   WaterAid, Water and Sanitation for the Urban Poor, and UNICEF. 2018. *Female-Friendly Public and Community Toilets: A Guide for Planners and Decision Makers.*

[14]   S.M. Hartigan et al. 2020. Why Do Women Not Use the Bathroom? Women's Attitudes and Beliefs on Using Public Restrooms. *International Journal of Environmental Research and Public Health.* 17 (6).

[15]   N. Abrahams, S. Mathews, and P. Ramela. 2006. Intersections of 'Sanitation, Sexual Coercion and Girls' Safety in Schools.' *Tropical Medicine and International Health.* 11 (5). pp. 751–756.

[16]   UNICEF and WHO. 2023. *Progress on Household Drinking Water, Sanitation and Hygiene, 2000–2022: Special Focus on Gender.*

[17]   United Nations Economic and Social Commission for Asia and the Pacific (UNESCAP). 2022. *A Three-Decade Journey Towards Inclusion: Assessing the State of Disability-Inclusive Development in Asia and the Pacific.*

[18]   CBM Australia and Water for Women. 2020. *Disability Inclusion and COVID-19: Guidance for WASH Delivery.*

[19]   World Economic Forum. The Valuable 500—Closing the Disability Inclusion Gap. https://www.weforum.org/projects/closing-the-disability-inclusion-gap.

Some people with disabilities also experience additional forms of discrimination based on other aspects of their identity. For example, women and girls with disabilities face unique challenges when accessing public WASH facilities, especially in managing their menstrual needs.

It is important to note that different types of disabilities require different solutions. For example, people with physical disabilities need different enabling designs within public WASH facilities than those with visual impairments. Full inclusion of people with disabilities means addressing all forms of disabilities when designing or improving public WASH facilities.

People with Diverse Sexual Orientation, Gender Identity, Gender Expression, and Sex Characteristics

People with diverse sexual orientation, gender identity, gender expression, and sex characteristics (SOGIESC)—also referred to as LGBTQ+ or sexual and gender minorities—often experience challenges in accessing public WASH facilities due to legal barriers and social prejudice, resulting in discrimination, marginalization, and exclusion from essential services. For example, *hijra* and *kinnar* (transgender, intersex, and third-gender people) in Bangladesh and India report discrimination when accessing public water points because of stereotypes of uncleanliness,[20] while research from the United States shows that after disasters, people with diverse SOGIESC are more likely to be displaced and experience water insecurity and unsanitary conditions than heterosexual and cisgender people.[21]

Intersectionality is crucial when considering the needs of people with diverse SOGIESC, as they often face additional layers of discrimination due to other aspects of their identity. For example, women who are lesbian, bisexual, transgender, or intersex may face challenges not only because of their gender identity and sexual orientation but also because they are women. Other factors, such as health status (e.g., HIV/AIDS) and disability, further exacerbate exclusion in accessing public WASH facilities, both due to inappropriate infrastructure and discriminatory social attitudes.

Older People

People aged 65 and above now represent 10% of the world's population, a proportion that will grow as life expectancies increase, especially in developing Asia and the Pacific. By 2050, one in six people globally will be 65 or older, doubling the number of older people from 761 million in 2021 to 1.6 billion. East Asia, Southeast Asia, South Asia, and Central Asia will account for 60% of this increase.[22]

---

[20]   Water for Women, Edge Effect, and Australian Aid. 2020. *Sexual and Gender Minorities and COVID-19: Guidance for WASH Delivery*. Guidance note.

[21]   J. Geiger, M. Méndez, and L. Goldsmith. 2023. *Amplified Harm: LGBTQ+ Disaster Displacement*. University of California, Irvine.

[22]   UN Department of Economic and Social Affairs. 2023. *World Social Report 2023: Leaving No One Behind in an Ageing World*.

As people age, they encounter new difficulties in accessing WASH facilities and looking after their personal needs becomes essential, particularly their increased need to use toilets. Public WASH facilities must respond to these needs, especially in terms of physical design and assistive devices. Components such as handrails beside toilets and handwashing sinks, motion-activated or easy-to-turn taps, and visual aids can assist older people with movement limitations and remind them to perform hygiene practices like flushing toilets and washing hands. Features designed for people with disabilities, such as ramps, also benefit older people.

People and Households Experiencing Poverty

Low-income countries, communities, and households have significantly less access to WASH than those with high incomes. Most households lacking access to safe drinking water and sanitation are impoverished, with cost being a major barrier. Even when access exists, services are often less reliable or of lower quality.[23] For example, in Fiji, only 29% of the poorest quintile has access to contamination-free water, compared with 82% of the richest quintile. In addition, rural areas globally have far less coverage of safely managed drinking water than urban areas (footnote 16).

The link between lower incomes and substandard WASH services also extends to public WASH facilities, especially since the majority of people in low- and middle-income countries rely on sanitation facilities shared by more than one household. Studies show that shared sanitation facilities and other public WASH facilities in low-income areas are generally of poorer quality, dirtier, less functional, and often lack strong operation and maintenance (O&M) arrangements.[24]

---

[23]   UNICEF and WHO. 2021. *The Measurement and Monitoring of Water Supply, Sanitation and Hygiene (WASH) Affordability: A Missing Element of Monitoring of Sustainable Development Goal (SDG) Targets 6.1 and 6.2.*

[24]   See, for example, P. Antwi-Agyei et al. 2020. Understanding the Barriers and Opportunities for Effective Management of Shared Sanitation in Low-Income Settlements—The Case of Kumasi, Ghana. *International Journal of Environmental Research and Public Health.* 17 (12); and M. Heijnen et al. 2015. Shared Sanitation Versus Individual Household Latrines in Urban Slums: A Cross-Sectional Study in Orissa, India. *American Journal of Tropical Medicine and Hygiene.* 93 (2). pp. 263–268.

Overview

WASH facilities are essential for applying a GESI approach. With a set of key principles, existing WASH facilities can be made inclusive, and new facilities can be designed to ensure access for all.

Gender- and socially inclusive WASH means understanding GESI's impact on access to WASH facilities and taking steps to overcome challenges, ensuring access for all. It implies that facilities are designed and constructed with GESI-responsive features to enable everyone—especially women, girls, people with disabilities, older people, and people with diverse SOGIESC—to use facilities safely, hygienically, and with dignity.

The need for inclusive WASH facilities goes beyond the household. In public buildings and spaces, inclusive WASH facilities remove barriers to participation. Poor-quality WASH facilities negatively impact how, when, and if women, girls, people with disabilities, and people with diverse SOGIESC access and use public buildings. For example, high rates of absenteeism and dropout in schools are often linked to inadequate WASH facilities that do not meet the needs of girls and children with disabilities. Schoolgirls in Fiji, for instance, report that they cannot concentrate in class because they worry about staining their uniforms when on their period and often go home early because they cannot change their menstrual materials at school.[25] In health facilities, unsuitable WASH facilities lead to high rates of infection and mortality, especially for mothers and children. Similarly, in government buildings that provide essential services, marginalized citizens may choose not to access these services out of fear of being unable to use the toilet when necessary.

The key principles of safely managed sanitation also apply to inclusive WASH facilities. These facilities must use flush or pour-flush toilets (with sufficient water for flushing) or squat latrines with slabs or platforms, and human waste must be safely disposed of, whether through connection to piped sewerage, on-site storage, or on-site or off-site treatment.

Gender- and socially inclusive WASH facilities should also be hygienic. This means ensuring users can wash their hands with water and soap, facilities are kept clean and dry, toilet paper and/or water is provided for personal cleaning, and waste disposal mechanisms are in place and well-managed (including regular and safe emptying). Where resources allow, free and easy-to-access menstrual materials such as sanitary pads should also be provided.

---

[25] UNICEF. 2017. *WASH in Schools Empowers Girls' Education in Fiji: An Assessment of Menstrual Hygiene Management in Schools.*

Investing in inclusive WASH facilities in public buildings will improve mobility, civic participation, and income generation. This, in turn, leads to greater equality and reduces discrimination based on gender, disability, and social identity. Access to sanitation is a basic right for everyone, and proper investments must be made to guarantee that inclusive WASH facilities are available in public spaces and buildings.

Key Principles

Existing national standards, guidelines, and policies on public WASH provide the primary reference points for public WASH facility design and improvement and should be followed whenever possible. Many constitutions in Asia and the Pacific, as well as international treaties such as the International Covenant on Economic, Social, and Cultural Rights, recognize citizens' rights to water and sanitation. The United Nations (UN) Committee on Economic, Social, and Cultural Rights acknowledges water as a human right under Article 11 on the right to an adequate standard of living and Article 12 on the right to the highest attainable standard of health.[26]

Recognizing human right to water and sanitation adds momentum to addressing GESI challenges. The human rights-based approach (HRBA) protects human dignity and advocates for equal access to water and sanitation for marginalized and vulnerable groups, empowering everyone to exercise their rights (Box 3). By incorporating HRBA principles and GESI considerations into the design, planning, and operation of WASH projects, inclusivity is strengthened, sustainable outcomes are achieved, and human dignity is upheld.[27]

Approaches to making WASH facilities inclusive are rapidly evolving, and much more can be done to ensure that public WASH facilities meet the needs of all users. This includes addressing what makes a toilet "usable," which depends on three factors: availability, functionality, and privacy (Box 4). Public WASH facilities must, therefore, consider and respond to the needs of women, children, people with disabilities, and people with diverse SOGIESC.

At a minimum, public WASH facilities should

   (i)   provide gender-segregated and gender-neutral facilities;
   (ii)   ensure safety and privacy;
   (iii)   provide water for toileting, washing, and handwashing;
   (iv)   incorporate disability and universal design features;
   (v)   support menstrual management; and
   (vi)   meet the needs of parents and caregivers.

---

[26]   UN Economic and Social Council. 2003. *General Comment No. 15: The Right to Water (Arts 11 and 12 of the Covenant).* Adopted at the 29th Session of the Committee on Economic, Social and Cultural Rights. 20 January.

[27]   ADB. 2024. *Paving the Inclusive Path Toward Water for All: Policies and Cases Supporting a Human Rights-Based Approach in Asia and the Pacific.* https://www.adb.org/publications/paving-inclusive-path-water-all.

Box 3: Human Rights-Based Approach and Human Right to Water and Sanitation

The following standards should be met to fulfill the human right to water and sanitation:

- **Accessibility.** Water and sanitation facilities and services must be accessible to everyone, without discrimination.
- **Availability.** There should be a sufficient and continuous water supply for personal and domestic use, including drinking, food preparation, and personal hygiene.
- **Affordability.** Water and sanitation services must be affordable for the population, ensuring no one is excluded based on cost.
- **Quality.** Water should be safe and free from contaminants that pose risks to human health.
- **Acceptability.** Water and sanitation facilities and services should be safe and private, allowing people to maintain dignity while also being culturally and socially acceptable.

Integrating the human right to water and sanitation with the human rights-based approach promotes a shift from a needs-based approach to a more holistic process. This alignment helps achieve Sustainable Development Goal 6.

Source: ADB. 2024. *Paving the Inclusive Path Toward Water for All: Policies and Cases Supporting a Human Rights-Based Approach in Asia and the Pacific.*

Box 4: What Is a "Usable" Toilet?

Toilets must be usable to meet users' sanitation needs. A usable toilet is defined by three key factors: availability, functionality, and privacy. These factors ensure that toilets serve their intended purpose:

- Toilets are not usable if they are inaccessible or locked, making them unavailable.
- Toilets are not usable if they are broken, blocked, overflowing, or lacking water, preventing proper functioning.
- Toilets are not usable if they lack privacy, such as missing lockable doors or having gaps in the walls.

Source: WHO and UNICEF. 2022. *Progress on WASH in Health Care Facilities, 2000–2021: Special Focus on WASH and Infection Prevention and Control.*

Gender-Segregated and Gender-Neutral Facilities

All facilities should be gender segregated, either in separate blocks (such as in schools) or with separate entrances, and at least one gender-neutral toilet should be provided. A larger number of toilets should be allocated for women and girls, as studies show that men spend an average of 60 seconds in the toilet, while women take 90 seconds (50% more time due to biological reasons like menstruation and social responsibilities such as caregiving. Many countries' national building codes and WASH policies often provide recommended ratios for male, female, gender-neutral, and accessible toilets. Examples from Nepal and Fiji are detailed in Section 4 and Section 5.[28]

Unisex or gender-neutral facilities are important as they can be used by parents with children, as well as by people who are transgender, nonbinary, third gender, or intersex. If space is limited, the gender-neutral toilet can also serve as a dedicated toilet for people with disabilities. Facilities should install clear signposts in local languages and use easily understood words or symbols to indicate male, female, gender-neutral, and disability-friendly facilities.

Safety and Privacy

Facilities should be located in easily accessible areas and not hidden away to reduce the risk of gender-based violence. Proper lighting is essential outside the entrance, inside the facility, and in each cubicle. Cubicles should be easily lockable from the inside, including by children and people with disabilities.

Water and Handwashing

All facilities should be equipped with handwashing sinks or taps, complete with soap and hand-drying options such as paper towels, cloth towels, or electric hand driers. For users of reusable menstrual materials, such as cloths, cloth pads, or menstrual cups, it is beneficial for some cubicles to have a sink and soap for both handwashing and washing menstrual materials for immediate or later reuse. While communal sinks are available for washing menstrual materials, users may avoid them due to the potential mess and embarrassment.

Universal Design

Universal design ensures that public WASH facilities are constructed to allow individuals of all identities and abilities to access and use the facilities safely, hygienically, and with dignity. Figure 6 shows a visual example of how these principles can be implemented.

For people with disabilities, including those using wheelchairs or other ambulatory devices, universal design integrates support features like ramps and handrails. Western flush toilets are generally preferred over squat toilets, which may be difficult for some people with disabilities. In public buildings with multiple toilet cubicles, at least one should be designed for wheelchair users. This cubicle should feature a door that is at least 80 centimeters wide and opens outward; sufficient space inside to maneuver (at least 1.5 meters wide by 2.2 meters deep); and features such as toilet, sink, and door lock that are at an appropriate height. If space is limited, this cubicle can also serve as a gender-neutral or unisex toilet.

---

28   Ghent University. 2017. No More Queuing at the Ladies' Room: How Transgender-Friendliness May Help in Battling Female-Unfriendly Toilet Culture. *ScienceDaily*. 14 July. www.sciencedaily.com/releases/2017/07/170714142749.htm.

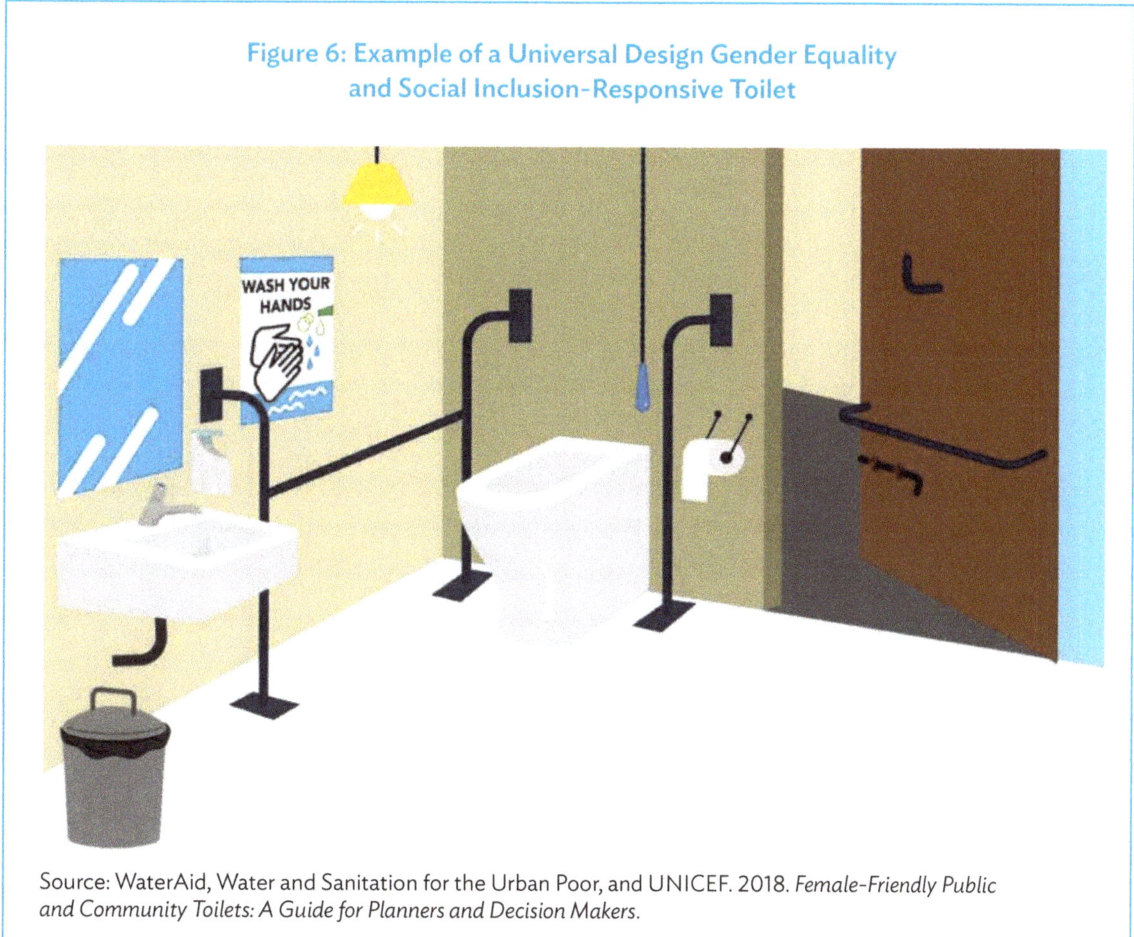

Figure 6: Example of a Universal Design Gender Equality and Social Inclusion-Responsive Toilet

Source: WaterAid, Water and Sanitation for the Urban Poor, and UNICEF. 2018. *Female-Friendly Public and Community Toilets: A Guide for Planners and Decision Makers.*

Children's needs must also be considered when designing WASH facilities. Toilets should be safe and easy for children, with door locks and handwashing stations positioned at child-friendly heights. In public buildings, it is important to ensure that at least one handwashing basin in both men's and women's toilets is accessible to children.

Menstrual Health and Hygiene

A major challenge in menstrual health is the cleaning and disposal of used menstrual materials, especially single-use sanitary pads and reusable cloths. This issue is exacerbated in areas with limited solid waste management. Menstrual waste includes used sanitary pads, tampons, cloths, and other materials intended to absorb menstrual blood (such as diapers), which are classified as solid waste. Safe management of menstrual waste involves treating and disposing of used menstrual materials to prevent harm to both people and the environment, including water systems.

Menstruation stigma has led to a global preference for private and straightforward disposal of menstrual products. With suitable facilities, people can avoid disposing of these materials in inappropriate places, such as rivers, fields, behind buildings (including schools and dormitories), or directly into toilets. These practices contribute to environmental pollution, blockages in toilets and sewerage systems, and unsanitary conditions. For instance, research in Bihar, India revealed that 60% of women discarded their menstrual waste outdoors, even when using reusable materials, to avoid keeping used menstrual products in their homes.[29]

Therefore, all toilet facilities must be equipped with proper means for cleaning and disposing of menstrual materials. These can range from simple solutions like sinks for washing reusable materials and bins for disposal, as illustrated in Figure 7, which shows an example of a gender-responsive toilet, to more sophisticated systems like incinerators, eco-friendly recycling processes, and even AI-powered disposal facilities. Ideally, disposal facilities should be available within each toilet cubicle for discreet disposal of materials and included in male and gender-neutral toilets to accommodate everyone who menstruates.

Effective disposal designs vary and are especially important in key locations such as schools and health facilities. Regular maintenance, including the emptying and cleaning of bins, is essential, while options like incinerators and recyclers should be environmentally considerate and well-maintained. Direct chutes from cubicles to incinerators are becoming increasingly popular, especially in schools and areas with inadequate waste collection services.

Parents and Caregivers

In much of the world, women and girls predominantly undertake caregiving roles, including for children, people with disabilities, and older or unwell family members, not only at home but also in public spaces. Public WASH facilities must address these needs by ensuring toilet cubicles are spacious enough for caregivers and those they are assisting, and by providing clean, safe areas for changing diapers.

Being able to change and dispose of diapers hygienically and safely is crucial, as improper disposal and insufficient handwashing facilities can spread illnesses.

Public WASH facilities should include baby changing stations in all toilets to foster inclusivity, recognizing that caregiving responsibilities are shared across genders. Ideally, these stations should be located in a dedicated "parents' room" outside the main toilet areas to enhance comfort, safety, hygiene, and accessibility for all caregivers.

---

[29]   Water Supply and Sanitation Collaborative Council. 2013. *Celebrating Womanhood: How Better Menstrual Hygiene Management is the Path to Better Health, Dignity and Business.*

Figure 7: Example of a Gender-Responsive Toilet

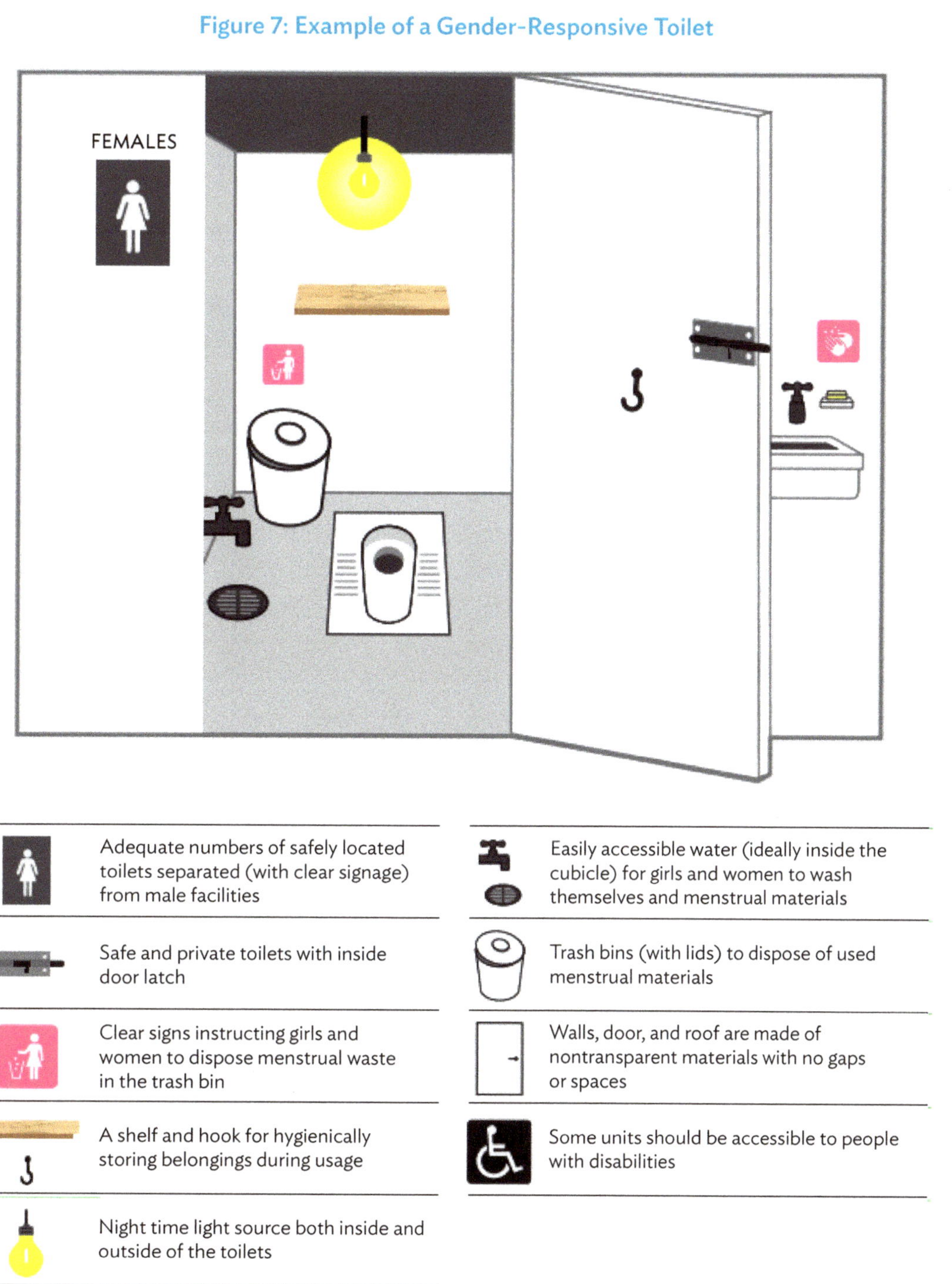

	Adequate numbers of safely located toilets separated (with clear signage) from male facilities		Easily accessible water (ideally inside the cubicle) for girls and women to wash themselves and menstrual materials
	Safe and private toilets with inside door latch		Trash bins (with lids) to dispose of used menstrual materials
	Clear signs instructing girls and women to dispose menstrual waste in the trash bin		Walls, door, and roof are made of nontransparent materials with no gaps or spaces
	A shelf and hook for hygienically storing belongings during usage		Some units should be accessible to people with disabilities
	Night time light source both inside and outside of the toilets		

Source: Columbia University and International Rescue Committee. 2017. *A Toolkit for Integrating Menstrual Hygiene Management (MHM) into Humanitarian Response.*

Summary of Key Components for Inclusive Water, Sanitation, and Hygiene Facilities

Recognizing the diverse needs of users is crucial in developing and improving public WASH facilities. Table 1 summarizes the key components to consider in ensuring all facilities are gender- and socially inclusive.

Table 1: Key Components for Gender- and Socially Inclusive Public Water, Sanitation, and Hygiene Facilities

Component	Principles	Solutions
Gender-segregated and gender-neutral facilities	1. All genders can safely access and use facilities with dignity and without discrimination. 2. Different biological needs and social roles should be recognized.	**Essential:** 1. Clearly labeled[a] gender-segregated facilities, with entrances located at a sufficient distance[b] apart 2. Larger number of female toilets than male toilets and urinals (ratio depends on local guidelines and context, but a 2:1 ratio of female to male toilets is a good guideline) **Desirable:** 1. Facilities for different genders completely separated (no shared walls) 2. Dedicated unisex or gender-neutral toilets
Safety and privacy	1. All genders can safely and privately access and use facilities without risk of violence or harassment, including at night.	**Essential:** 1. Facilities appropriately located: easily accessible and not hidden away or in locations perceived by users as dangerous 2. Adequate lighting: outside the entrance, inside the facility, and inside cubicles 3. Strong and easily lockable cubicle doors **Desirable:** 1. Trained caretakers of all genders always present (in their respective gender toilet sections) in large or busy public WASH facilities (e.g., in parks or town squares) 2. Caretakers trained to assist in emergencies (including in cases of sexual harassment and violence) 3. Security cameras (CCTV) at toilet block entrances (if appropriate)
Water and handwashing	1. Safely managed water available within facilities for all genders. 2. All genders have access to sufficient water and soap.	**Essential:** 1. Handwashing facilities with soap (inside or outside cubicles): sinks, taps, or buckets with scoops **Desirable:** 1. Additional washing facilities inside cubicles for menstruating individuals 2. Lockable shower rooms 3. Facilities for personal washing needs, such as *wudhu* or other forms of pre-prayer ablutions

continued on next page

Table 1 *continued*

Component	Principles	Solutions
Disability and universal design	1. People with disabilities of all genders can access and use toilets safely, hygienically, and with dignity. 2. Wheelchair and ambulatory device users can easily and safely access toilet buildings and cubicles. 3. Wheelchair and ambulatory device users can easily and safely move around inside toilet cubicles.	**Essential:** 1. Clear and globally recognizable symbols for male, female, and accessible toilets, with arrows and text in local language(s) 2. Disability-friendly access: paths (at least 1.2 meters wide), ramps (gradient of no more than 1 in 15), handrails, no steps 3. Disability-friendly cubicles: wide outward-opening doors (minimum width of 80 centimeters), sufficient space (1.5 meters wide by 2.2 meters deep), handrails, accessible sinks and taps, dry floors, large easy-to-use door locks 4. Tactile paving and signs for the visually impaired **Desirable:** 1. Child-friendly design in all facilities, such as smaller toilets and lower sinks 2. Trained caretakers of all genders to assist people with disabilities
Menstrual health and hygiene	1. People who menstruate can change their menstrual materials safely, hygienically, and with dignity. 2. People who menstruate can dispose of used single-use menstrual materials hygienically and appropriately. 3. People who menstruate can wash themselves with water. 4. People who menstruate can wash reusable menstrual materials.	**Essential:** 1. Safe and hygienic disposal mechanisms for used menstrual materials (e.g., regularly cleaned bins or buckets with lids) **Desirable:** 1. Water for managing menstruation inside cubicles: sink, tap, or bucket with scoop, plus drain 2. Hooks and ledges for belongings 3. Easy-to-access and free or low-cost menstrual products
Parents and caregivers	1. Parents and caregivers can safely and hygienically care for others. 2. Parents, caregivers, and those they care for have sufficient space to comfortably manage both their own hygiene needs and those of the person or child in their care.	**Essential:** 1. Baby changing facilities in all toilets (not just female toilets) 2. Safe and hygienic disposal mechanisms for diapers and wipes, regularly emptied and cleaned 3. Sufficient space inside cubicles for parents and caregivers 4. Separate breastfeeding or parenting rooms

[a] Clearly labeled facilities should use internationally recognized symbols as well as writing or other signs that are easily understood by local communities to indicate gender and disabled access.
[b] Local communities should be consulted to determine an appropriate and safe distance between toilet blocks and toilet entrances for different genders.

Source: Asian Development Bank.

Checklist 1 can be used to assess whether an existing public WASH facility is GESI-responsive before identifying necessary improvements. It can also serve as a checklist when designing new public WASH facilities.

Checklist 1: Assessing the Gender Equality and Social Inclusion Status of Public Water, Sanitation, and Hygiene Facilities

Component	Status		
Gender-Segregated and Gender-Neutral Facilities			
Essential			
Clearly labeled[a] gender-segregated facilities	Yes	No	
Facility entrances for different genders located at a sufficient distance[b] apart	Yes	No	
Greater number of female toilets than male toilets and urinals (in line with national ratio, if available)	Yes	No	
Desirable			
Facilities for different genders completely separated (no shared walls)	Yes	No	
Dedicated unisex or gender-neutral toilet(s)	Yes	No	
Safety and Privacy			
Essential			
Facilities appropriately located: easily accessible and not hidden away or in locations perceived by users as dangerous	Yes	No	
Adequate lighting: outside the entrance, inside the facility, and inside cubicles	Yes, functioning	Yes, not functioning	Not present
Strong and easily lockable cubicle doors	Yes, functioning	Yes, not functioning	Not present
Desirable			
Trained caretakers of all genders always present (in their respective gender toilet sections) in large or busy public WASH facilities (e.g., in parks or town squares)	Yes	No	
Caretakers trained to assist in emergencies (including in cases of sexual harassment and violence)	Yes	No	
Security cameras (CCTV) at toilet block entrances	Yes, functioning	Yes, not functioning	Not present

continued on next page

Checklist 1 *continued*

Component	Status		
Water and Handwashing			
Essential			
Safely managed water available within facilities for all genders	Yes		Not present
Handwashing facilities with soap (located inside or outside cubicles): sinks, taps, or buckets with scoops	Yes, functioning	Yes, not functioning	Not present
Desirable			
Additional washing facilities inside cubicles for menstruating individuals	Yes, functioning	Yes, not functioning	Not present
Facilities for showering (in dedicated lockable shower rooms)	Yes, functioning	Yes, not functioning	Not present
Facilities for other personal washing needs, such as *wudhu,* a ritual washing performed by Muslims before prayer, and other ritual cleansing purposes	Yes, functioning	Yes, not functioning	Not present
Disability and Universal Design			
Essential			
Clear and globally recognizable symbols for male, female, and accessible toilets, with arrows and text in local language(s)	Yes, intact	Yes, damaged	Not present
Tactile paving and signs for the visually impaired	Yes, intact	Yes, damaged	Not present
Disability-friendly access: paths (at least 1.2 meters wide), ramps (gradient of no more than 1 in 15), handrails, no steps	Yes		No
Disability-friendly cubicles: wide outward-opening doors (minimum width of 80 centimeters), sufficient space (1.5 meters wide by 2.2 meters deep), handrails, accessible sinks and taps, dry floors, large easy-to-use door locks	Yes		No
Desirable			
Child-friendly design in all facilities, such as smaller toilets and lower sinks	Yes, functioning	Yes, not functioning	Not present
Trained caretakers of all genders to assist people with disabilities	Yes		No
Menstrual Health and Hygiene			
Essential			
Safe and hygienic disposal mechanism for used menstrual materials (e.g., regularly cleaned bins or buckets with lids)	Yes, functioning	Yes, not functioning	Not present

continued on next page

Checklist 1 *continued*

Component	Status		
Desirable			
Water for managing menstruation inside cubicles: sink, tap, or bucket with scoop, plus drain	Yes, functioning	Yes, not functioning	Not present
Hooks and ledges for belongings	Yes, intact	Yes, damaged	Not present
Easy-to-access and free or low-cost menstrual products	Yes		No
Parents and Caregivers			
Essential			
Baby changing facilities in all toilets (not just female toilets)	Yes, intact	Yes, damaged	Not present
Safe and hygienic disposal mechanisms for diapers and wipes, regularly emptied and cleaned	Yes, intact	Yes, damaged	Not present
Sufficient space inside cubicles	Yes		No
Desirable			
Baby changing facilities in toilets for all genders	Yes, intact	Yes, damaged	Not present

[a] Clearly labeled facilities should use internationally recognized symbols as well as writing or other signs that are easily understood by local communities to indicate gender and disabled access.
[b] Local communities should be consulted to determine an appropriate and safe distance between toilet blocks and toilet entrances for different genders.

Source: Asian Development Bank.

While Checklist 1 can be used to directly assess public WASH facilities, Appendix 1 provides a checklist for evaluating national and subnational enabling environments, covering WASH sector policies, budgets, and plans.

Inclusive Water, Sanitation, and Hygiene in Health Care Facilities

Safely managed water and sanitation services in health care facilities significantly improve hygiene, preventing infections and the spread of communicable diseases. This is especially critical for pregnant women, women in childbirth, and newborns. WASH and infection prevention are closely linked; the WHO/UNICEF Joint Monitoring Programme for Water Supply, Sanitation and Hygiene (JMP) describes the two components as "interdependent and complementary."[30]

---

[30]   WHO and UNICEF. 2022. *Progress on WASH in Health Care Facilities, 2000–2021: Special Focus on WASH and Infection Prevention and Control (IPC).* p. 16.

To improve infection prevention and control, both the UN and WHO have called for increased action to improve WASH in health care facilities. In 2018, UN Secretary-General António Guterres launched a global call for action on WASH in health care facilities:

> "A recent survey of 100,000 [health care] facilities found that more than half lack simple necessities, such as running water and soap—and they are supposed to be healthcare facilities. The result is more infections, prolonged hospital stays and sometimes death. We must work to prevent the spread of disease. Improved water, sanitation and hygiene in health facilities is critical to this effort."[31]

The JMP tracks five essential services for WASH delivery in health care facilities: water, sanitation, hygiene, waste management, and environmental cleaning (Table 2). The service ladders define three levels of service—no service, limited service, and basic service—and note that in countries where basic service is already standard, a country-defined advanced service level should be the target.

Many countries have existing standards or guidelines for health care facilities. In Asia and the Pacific, examples include Cambodia (2018 National Guidelines for Water, Sanitation and Hygiene in Health Care Facilities); Fiji (2020 Guidelines for Climate-Resilient and Environmentally Sustainable Health Care Facilities in Fiji, see Section 5 for a case study of Fiji); Nepal (2018 National Standards for WASH in Health Care Facilities, see Section 4 for a case study of Nepal); and Mongolia (2016 Environmental Hygienic Requirements for Health Facilities). Some countries in the Pacific, such as Papua New Guinea, have developed national policies based on the Australasian Health Facility Guidelines as primary references, supplemented with customized standard components for the national context.

Countries without national standards or guidelines covering all elements of WASH in health care facilities should strongly consider developing and implementing them.

Existing national standards and guidelines should be used as first points of reference for designing and improving WASH in health care facilities. Global standards, such as those defined by WHO, are also useful and should be applied alongside national policy documents. This is especially important in emerging health problems, like the coronavirus disease 2019 (COVID-19) pandemic, where standards may need updating. WHO's Essential Environmental Health Standards in Health Care (2008) is an excellent reference for improving WASH in health care facilities.

Checklist 1 can be used as an initial guide to assess existing WASH infrastructure in health care facilities. Additional criteria can be integrated based on national guidelines and standards or the WHO and UNICEF Water and Sanitation for Health Facility Improvement Tool (WASH FIT) (2nd edition, 2022), a risk-based quality improvement tool for health care facilities.

---

[31]   UN. 2018. Secretary-General's Remarks at Launch of International Decade for Action "Water for Sustainable Development" 2018–2028 [as Delivered]. 22 March. https://www.un.org/sg/en/content/sg/statement/2018-03-22/secretary-generals-remarks-launch-international-decade-action-water.

Table 2: Water, Sanitation, and Hygiene in Health Care Facilities Service Ladders

Service Level	Water	Sanitation	Hygiene	Waste Management	Environmental Cleaning
Basic Service	Water is available from an improved source* on the premises.	Improved sanitation facilities* are usable, with at least one toilet dedicated for staff, at least one sex-separated toilet with menstrual hygiene facilities, and at least one toilet accessible for people with limited mobility.	Functional hand hygiene facilities (with water and soap and/or alcohol-based hand rub) are available at points of care, and within 5 meters of toilets.	Waste is safely segregated into at least three bins, and sharps and infectious waste are treated and disposed of safely.	Protocols for cleaning are available, and staff with cleaning responsibilities have all received training.
Limited Service	An improved water source is available within 500 meters of the premises, but not all requirements for a basic service are met.	At least one improved sanitation facility is available, but not all requirements for a basic service are met.	Functional hand hygiene facilities are available either at points of care or toilets but not both.	There is limited separation and/or treatment and disposal of sharps and infectious waste, but not all requirements for a basic service are met.	There are cleaning protocols and/or at least some staff have received training on cleaning.
No Service	Water is taken from unprotected dug wells or springs, or surface water sources; or an improved source that is more than 500 meters from the premises; or there is no water source.	Toilet facilities are unimproved (e.g., pit latrines without a slab or platform, hanging latrines, bucket latrines) or there are no toilets.	No functional hand hygiene facilities are available either at points of care or toilets.	There are no separate bins for sharps or infectious waste, and sharps and/or infectious waste are not treated/disposed of.	No cleaning protocols are available, and no staff have received training on cleaning.

* Improved water sources are those that by the nature of their design and construction have the potential to deliver safe water. These include piped water, boreholes or tubewells, protected dug wells, protected springs, rainwater, and packaged or delivered water. Improved sanitation facilities are those designed to hygienically separate human excreta from human contact. These include wet sanitation technologies such as flush and pour-flush toilets connecting to sewers, septic tanks or pit latrines; and dry sanitation technologies such as dry pit latrines with slabs and composting toilets.

Source: World Health Organization (WHO) and United Nations Children's Fund (UNICEF). 2022. *Progress on WASH in Health Care Facilities, 2000–2021: Special Focus on WASH and Infection Prevention and Control (IPC)*.

Inclusive Water, Sanitation, and Hygiene in Schools

Gender- and socially inclusive WASH facilities are critical in schools. Governments should ensure all schools have functioning, child-friendly, and gender-separated toilets with menstrual hygiene management facilities. Children spend 7 hours or more a day in school, and inadequate WASH facilities—especially for girls and children with disabilities—can hinder attendance and academic performance. Additionally, close classroom environments increase the risk of disease transmission, making proper hygiene essential.

In general, national building codes and guidelines for public buildings provide sound guidance on minimum WASH standards, and countries in Asia and the Pacific often have dedicated WASH policies and procedures for schools. These and the JMP service ladders for WASH in schools should be used as a starting point (Table 3).

Table 3: Water, Sanitation, and Hygiene in Schools Service Ladders

Service Level	Drinking Water	Sanitation	Hygiene
Basic Service	Drinking water from an improved source and water is available at the school at the time of the survey	Improved sanitation facilities at the school that are single-sex and usable (available, functional, and private) at the time of the survey	Handwashing facilities with water and soap available at the school at the time of the survey
Limited Service	Drinking water from an improved source but water is unavailable at the school at the time of the survey	Improved sanitation facilities at the school that are either not single-sex or not usable at the time of the survey	Handwashing facilities with water but no soap available at the school at the time of the survey
No Service	Drinking water from an unimproved source or no water source at the school	Unimproved sanitation facilities or no sanitation facilities at the school	No handwashing facilities or no water available at the school

Source: World Health Organization (WHO) and United Nations Children's Fund (UNICEF). 2022. *Progress on Drinking Water, Sanitation and Hygiene in Schools: 2000–2021 Data Update.*

Key principles for WASH in schools include the following:

(i)   Sufficient and reliable water supply for drinking, handwashing, and hygiene purposes;
(ii) Sufficient and separate toilets for boys and girls, with lockable doors and appropriate gender ratios (check national standards for these ratios or international guidelines from organizations such as UNICEF);
(iii) Open defecation-free school grounds and surrounding areas;
(iv) Toilets that meet the accessibility needs of students and staff with disabilities;
(v)   Safe and private spaces for changing menstrual materials in female toilets, ideally with washing facilities; and
(vi) Handwashing facilities near toilets and areas where students gather, with clean water, soap, and drying options (there should be sufficient handwashing facilities based on a ratio of facilities to students).

Three Star Approach

The Three Star Approach for WASH in Schools, developed by UNICEF and the Deutsche Gesellschaft für Internationale Zusammenarbeit, helps schools and government agencies achieve key WASH facility standards to promote healthy and clean learning environments. Launched in 2013, the approach is now widely adopted globally.

The approach encourages schools to implement simple, low-cost measures to ensure that students regularly wash their hands with soap, have access to drinking water, and use clean and gender-segregated toilets. It prioritizes essential actions to achieve hygiene goals, guiding users through a clear pathway from minimum provisions to two stars, and ultimately, three stars. A school attains a three-star designation when it meets national standards (Figure 8).

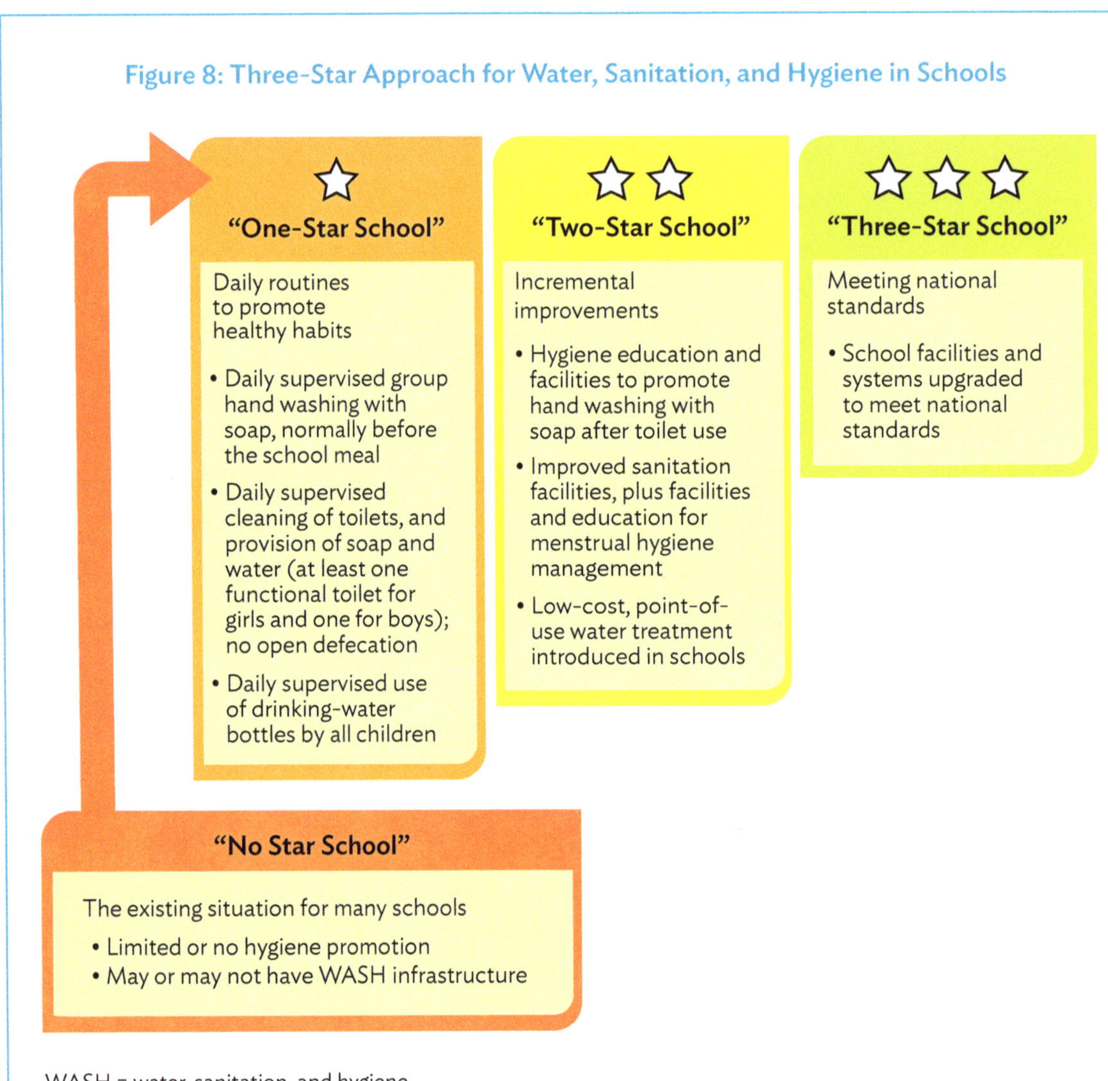

Figure 8: Three-Star Approach for Water, Sanitation, and Hygiene in Schools

WASH = water, sanitation, and hygiene.
Source: United Nations Children's Fund (UNICEF) and Deutsche Gesellschaft für Internationale Zusammenarbeit. 2013. *Field Guide: The Three Star Approach for WASH in Schools.*

Table 4 presents a simplified overview of the different star levels based on UNICEF's *Field Guide: Three Star Approach for WASH in Schools* (2013).

Table 4: Three-Star Approach for WASH in Schools

One-star schools: *Daily routines to promote healthy habits*

Criteria	Interventions	Results
All children participate in daily supervised group handwashing with soap, ideally before the school meal.	Daily supervised group handwashing with soap, ideally before meals	• Handwashing becomes a habit. • The need to wash hands before eating is reinforced. • Children learn proper handwashing techniques. • Group handwashing sessions provide a set time to deliver hygiene messages.
The school has basic gender-segregated toilets that are functional, clean, and used by all children (no open defecation).	Daily supervised cleaning and provisioning of toilets with soap and water	• Toilets are used because they are clean. • Water and soap are available in toilets. • Open defecation in and near the school is eliminated. • Children learn the importance of sanitation through active participation.
Every child has and correctly uses a personal drinking water bottle.	Children have access to drinking water at school whether or not a safe source is available on campus premises.	• Children have access to drinking water at school whether or not a safe source is available on campus premises.

Two-star schools: *Incremental improvements*

Criteria	Interventions	Results
Children wash their hands with soap after using the toilet.	Hygiene education expanded to stress handwashing after toilet use, handwashing stations installed as needed, and menstrual hygiene education delivered in schools	• Children learn to wash their hands with soap at critical times: before meals and after defecation. • Handwashing stations are demonstrated to the community. • Girls gain knowledge and support on menstrual hygiene management.
Improved sanitation and menstrual hygiene facilities are available.	Additional and/or improved toilets, plus facilities for menstrual hygiene management, constructed where needed	• Additional toilets are available at schools for boys and girls. • Girls are further encouraged to attend classes because there are additional private sanitation and/or menstrual hygiene management facilities.
Drinking water is available at school.	Low-cost, point-of-use water treatment introduced in schools	• Children have access to safe drinking water at school. • Low-cost water treatment is demonstrated to the community.

continued on next page

Table 4 *continued*

Three-star schools: *Meeting national standards*		
Criteria	**Interventions**	**Results**
Schools meet national standards for WASH.	School facilities and systems upgraded to meet national standards	• Social norms on good hygiene behavior are institutionalized. • The school can offer full accessibility to WASH for all students, including children with disabilities. • National inequities are eliminated by ensuring all schools in the country have the same standards for WASH.

WASH = water, sanitation, and hygiene.
Source: United Nations Children's Fund (UNICEF) and Deutsche Gesellschaft für Internationale Zusammenarbeit. 2013. *Field Guide: The Three Star Approach for WASH in Schools.*

Checklist 1 can be used as a guide to assess the existing WASH infrastructure in schools. Additional points can be integrated based on national guidelines and standards.

Community Engagement

In both improving existing WASH facilities and planning new ones, community engagement is key to ensuring they are GESI-responsive, especially in public spaces. It is essential to consult and actively involve women, children, people with disabilities, sexual and gender minorities, discriminated-against castes, and other marginalized groups. Local communities and potential WASH facility users should be prioritized in engagement.

Research shows that engaging local and marginalized groups in WASH facility design and implementation improves not only GESI responsiveness but also overall effectiveness. For example, a World Bank evaluation of 122 water projects found that projects were six to seven times more effective when women were involved in planning, implementation, and monitoring.[32] This is because women's involvement and use of water and sanitation facilities differ from men's, both at home and in public. In addition, local communities better understand how spaces are used, where water resources are located and how they are managed, where open defecation occurs, and how women and girls manage menstruation.[33]

---

[32]   J. Fisher. 2006. *For Her It's The Big Issue: Putting Women at the Centre of Water Supply, Sanitation and Hygiene.* Water, Sanitation and Hygiene: Evidence Report. Water Supply and Sanitation Collaborative Council.

[33]   S. Nelson et al. 2021 How Community Participation in Water and Sanitation Interventions Impacts Human Health, WASH Infrastructure and Service Longevity in Low-Income and Middle-Income Countries: A Realist Review. *BMJ Open.* 11 (12).

Involving women and other marginalized groups, particularly those using the facilities, throughout the process ensures that WASH facilities are appropriate and meet users' needs. Below are key points at which local and marginalized groups should be engaged:

(i)   during needs assessment, resource identification, and idea generation;
(ii)   during the development of policies, procedures, guidelines, and standards;
(iii)   during planning, design, and budget allocation for facilities and interventions;
(iv)   in construction and/or improvement of facilities;
(v)   in testing, trialing, and piloting of facilities and interventions;
(vi)   in the implementation of behavior change interventions;
(vii)   in the operation and maintenance (O&M) of facilities; and
(viii) in monitoring and evaluation activities.

By involving a wide range of stakeholders at all stages, WASH facilities in public buildings will better serve users, be more sustainable, and contribute to meeting national targets.

The following steps are recommended to ensure community engagement during the design of public WASH facilities. Not all steps need to be followed every time. For example, stakeholder mapping may already exist from recent activity, or discussions about site selection may not be possible because of limited options (e.g., if the facility is being constructed within a school or health care facility.)

Checklist 2: Community Engagement in Designing Public Water, Sanitation, and Hygiene Facilities

Step	Process	Implemented?	
1. Stakeholder mapping	Mapping all relevant stakeholders, including marginalized groups	Yes	No
2. Discussions with stakeholders	Holding discussions with stakeholders (ideally gender-disaggregated) to identify users' needs, facility locations, and facility design components	Yes	No
3. Selection of final location	Stakeholders participating in and agreeing on the facility's final location	Yes	No
4. Agreement on final design	Stakeholders agreeing on the final design, with repetition if the draft does not result in agreement	Yes	No
5. Discussions with stakeholders on construction and O&M	Discussing construction of the facility and O&M with stakeholders, including community involvement	Yes	No
6. Monitoring and evaluation	Actively involving stakeholders in the monitoring and evaluation of the facility	Yes	No

O&M = operation and maintenance.
Source: Asian Development Bank.

Operation and Maintenance of Water, Sanitation, and Hygiene Infrastructure

Designing and constructing gender- and socially inclusive public WASH facilities is not the final step. O&M are routine activities necessary to keep facilities functioning properly. Without a solid O&M system, public WASH facilities can become unhygienic, dilapidated, and even dangerous, discouraging use and reducing participation in public life, education, and health care.

Maintenance for WASH infrastructure includes preventative maintenance (regular work to ensure infrastructure stays in good condition), corrective maintenance (repair or replacement of parts), and reactive maintenance (work that reacts to failures, malfunctions, or breakdowns).

Effective O&M requires planning and coordination among relevant stakeholders, including the community, to clarify roles and responsibilities, ensuring all types of maintenance are addressed as needed. Responsible government agencies must consistently allocate budgets and provide both financial and human resources to sustain O&M activities and ensure the sustainability of WASH infrastructure.

Key O&M activities for public WASH facilities include

(i)   ensuring continued water supply,
(ii)   maintaining cleanliness and hygiene,
(iii) maintaining physical components (including repair and replacement as needed),
(iv) ensuring safe access,
(v)   desludging on-site sanitation systems, and
(vi) promoting proper hygiene and use of facilities.

To ensure proper O&M, the facility's agencies should undertake regular monitoring and evaluation activities and provide a complaints mechanism. Checklist 3 can be used to monitor O&M to keep WASH facilities functional and inclusive.

Checklist 3: Monitoring the Operation and Maintenance of Public Water, Sanitation, and Hygiene Facilities

Component	Status	
Enabling Environment		
Is there a budget allocated for the facility's operation and maintenance (O&M)?	Yes	No
If yes: Is the budget allocation sufficient?	Yes	No
If yes: Does the budget allocation increase each financial year?	Yes	No
Is there a standard operating procedure for the facility?	Yes	No
If yes: Is the procedure implemented as designed?		

continued on next page

Checklist 3 *continued*

Component	Status	
Is there a monitoring and evaluation plan for the facility?	Yes	No
If yes: Is the plan implemented as designed?	Yes	No
If yes: Are community members and facility users involved in monitoring and evaluation activities?	Yes	No
Is it clear who is responsible for funding the O&M of the facility? (e.g., a certain government agency or institutional committee)	Yes	No
Is it clear who is responsible for performing the O&M of the facility? (e.g., a certain government agency or institutional committee)	Yes	No
Is there a functioning complaints mechanism?	Yes	No
If yes: Is information about how to make a complaint displayed inside or outside the toilet blocks or sections?	Yes	No
If yes: Are complaints acted upon within an appropriate time frame?	Yes	No
Accessibility, Gender, and Social Inclusion		
Are the facilities gender-segregated?	Yes	No
Is there a functioning unisex or gender-neutral toilet?	Yes	No
Are there enough female, male, and unisex or gender-neutral toilets for the current facility users? (Refer to national standards; 2:1 is a good rule of thumb for female-to-male toilets)	Yes	No
Are functioning child-friendly toilets and handwashing facilities available?	Yes	No
Is the toilet block or section accessible to people with physical disabilities, including wheelchair users? (e.g., appropriate gradient ramp, no steps)	Yes	No
Is there a functioning accessible toilet cubicle for people with physical disabilities, including wheelchair users?	Yes	No
Are there enough accessible toilets? (1:20 is a good rule of thumb)	Yes	No
Are there design elements for people with visual impairments (e.g., tactile paving and braille)?	Yes	No
Are there rubbish bins for used menstrual products inside each cubicle:		
In the female toilet block or section?	Yes	No
In the male toilet block or section?	Yes	No
In the unisex or gender-neutral toilet block or section?	Yes	No
Are baby changing facilities available (changing table and rubbish bin):		
In the female toilet block or section?	Yes	No
In the male toilet block or section?	Yes	No
In the unisex or gender-neutral toilet block or section?	Yes	No

continued on next page

Checklist 3 *continued*

Component	Status	
Water and Hygiene		
Is water available at the facility?	Yes	No
Is there sufficient water for flushing the toilet?	Yes	No
Is there a functioning tap or sink inside each cubicle?	Yes	No
If yes: Is there sufficient water for personal washing (including anal cleansing) and/or washing of used menstrual products?	Yes	No
Is there a functioning bidet in each toilet for anal cleansing?	Yes	No
Is there a functioning handwashing sink inside the toilet block or room?	Yes	No
If yes: Is there sufficient water for handwashing?	Yes	No
If yes: Is there soap available for handwashing?	Yes	No
If toilet paper is provided: Is there toilet paper in all cubicles?	Yes	No
Is there a functioning hand-drying facility? (e.g., paper towel dispenser, electric hand dryer)	Yes	No
Cleanliness		
Are all toilets clean and free of urine and feces?	Yes	No
If urinals are present: Are all urinals clean and free of urine?	Yes	No
Are all handwashing sinks clean?	Yes	No
Is the floor clean, both inside and outside the cubicles?	Yes	No
Is the facility visually free of urine and feces?	Yes	No
Is the facility free of foul smells and odors?	Yes	No
Is the facility free of rubbish on floors and behind and around toilets?	Yes	No
Is there a rubbish bin near the handwashing sinks?	Yes	No
Are all rubbish bins regularly emptied?	Yes	No
Is there a person (or persons) responsible for cleaning the facility?	Yes	No
Is there a cleaning schedule?	Yes	No
If yes: Is the cleaning schedule followed?	Yes	No
In open areas, such as schools, health facilities, and parks: Is the area surrounding the toilet block free of open defecation?	Yes	No

continued on next page

Checklist 3 *continued*

Component	Status	
Maintenance		
Are all symbols for male, female, and accessible toilets clearly visible?	Yes	No
Are all toilet components (e.g., flush mechanism, toilet seat, cistern) functioning and not broken or damaged?	Yes	No
Are all handwashing components (e.g., tap, basin) functioning and not broken or damaged?	Yes	No
Are all cubicle doors functioning, lockable, and not broken or damaged?	Yes	No
Are the floors, walls, windows, and ceilings intact and not broken or damaged?	Yes	No
Are all water and sewerage pipes functioning and not broken or damaged?	Yes	No
If present, is the sewerage containment facility (e.g., cesspit, septic tank) functioning, not overflowing, and not broken or damaged?	Yes	No
If present, are all lights functioning and not broken or damaged?	Yes	No
If present, are all solar panels functioning and not broken or damaged?	Yes	No
If present, are all hand-drying facilities (e.g., electric hand dryer, paper towel dispenser) functioning and not broken or damaged?	Yes	No

Source: Asian Development Bank

4 CASE STUDY—NEPAL

Overview

Nepal has made significant progress in WASH access since 2000. In 2000, just 30% of the population used sanitation facilities, and 73% had access to improved drinking water sources.[34] By 2022, at least 80% had access to basic sanitation, and 51% had access to safely managed sanitation. Meanwhile, 91% had access to basic drinking water, although access to safely managed water dropped to 16% in 2023 from 25% in 2015 due to contamination of water sources after the April 2015 earthquake. A total of 64% of people have access to basic hygiene facilities, and open defecation has largely ceased, falling from 67% in 2000 to just 7% in 2022 (footnote 16), leading to the government's declaration of Nepal as open defecation-free in 2019.

However, although household access to water and sanitation has improved substantially, challenges remain for WASH facilities in public buildings. Health care facilities are a key example. The 2022 Joint Monitoring Programme (JMP) for Water Supply, Sanitation, and Hygiene statistics for WASH in health care facilities indicated that 6% of nonhospital health care facilities in Nepal had no access to water, 30% lacked piped water, and only 80% were "visibly clean." On sanitation, 12% of nonhospital facilities had either no sanitation services or only unimproved services.[35]

Schools also face WASH challenges. In the 2020–2021 school year, the JMP found that only 47% of schools in Nepal had basic water services (assessed as being "improved and available"), 33% had limited water services, and 21% had no water services. Secondary schools were better equipped, with 76% having basic water services compared to 39% of primary schools. Data on school sanitation services are limited, but the 2022 JMP data noted that only 9% of schools had no sanitation services.[36] A 2021 SNV survey found that 99% of schools in four cities in Nepal had toilets, but the quality varied significantly— 52% used unimproved toilets, 49% lacked separate toilets for boys and girls, and 10% lacked handwashing stations (of those that did, 38% lacked soap).[37]

Nepal's public toilet situation is also poor. A 2022 assessment of public toilets (including those in government buildings, hospitals, commercial buildings, religious buildings, and petrol stations) in the Kathmandu Valley found most in poor condition, with low scores from "okay" to "very bad" for gender friendliness, menstrual hygiene management, and disability access. The report recommended that local governments set standards for public toilets and develop O&M guidelines.[38]

---

34 Government of Nepal, National Planning Commission. 2017. *National Review of Sustainable Development Goals.*
35 WHO and UNICEF. 2022. *Progress on WASH in Health Care Facilities, 2000–2021: Special Focus on WASH and Infection Prevention and Control (IPC).*
36 UNICEF and WHO. 2022. *Progress on Drinking Water, Sanitation and Hygiene in Schools: 2000–2021 Data Update.*
37 SNV. 2021. *Standard Operating Procedures for Water, Sanitation, and Hygiene in Schools.*
38 WaterAid Nepal and GUTHI. 2022. *Mapping and Enumeration of Public Toilets: Status Report on Female-Friendly Public Toilets.*

How Gender and Social Identity Affect Access to Water, Sanitation, and Hygiene in Nepal

Children

Poor WASH access at schools affects not only attendance rates but also health. Research shows that children with poor access to WASH at school also fall ill more often. A study of schools in Chitwan and Dhanusha districts found that 64% of students from schools with unimproved WASH facilities reported being ill, compared with 41% of students from schools with improved facilities. Notably, gender and social identity also influenced illness rates: 57% of girls reported illness versus 47% of boys. Among students in schools with unimproved facilities, 67% of Dalit children and 62% of Brahmin/Chhetri-Terai children reported illness, compared with lower rates among Janajati (43%) and Brahmin/Chhetri-Hill (39%) students.[39] This demonstrates how gender and social identity shape the challenges individuals face in accessing WASH services and facilities.

Women and Girls

In Nepal, menstruation significantly restricts women's and girls' participation in public life. A 2019 study showed that, although 87% of women and girls aged 15–49 had access to a private place for washing and changing and 94% used menstrual materials (such as sanitary pads), less than 1% engaged in daily activities during menstruation due to high levels of stigma.[40]

"Of course, you couldn't fault them… The toilets in the school were in a bad condition, sanitary pads were not available, and there were no means of disposing used ones either."

– Prem Bahadur Thapa, Principal, Bhanu Secondary School, Simta, Surkhet District

Source: United Nations Children's Fund Nepal. 2022. Good for Everyone. https://www.unicef.org/nepal/stories/good-everyone.

For example, schoolgirls in Nepal may miss school or leave early while menstruating, partly due to social stigma, which views menstruation as "dirty," but also because school WASH facilities are inadequate for managing periods safely, comfortably, and hygienically. In 2021, only 41% of surveyed schools in Nepal had facilities for menstrual hygiene management, including covered bins, changing spaces, and water (footnote 38). Without these facilities, menstruating schoolgirls and staff cannot manage their hygiene effectively (e.g., changing their sanitary pads or washing themselves).

People with Disabilities

According to Nepal's 2014–2015 National Living Standards Survey, some 3.6% of the population lives with disabilities. Despite the Constitution of Nepal supporting disability inclusion, government WASH policies have been assessed as inadequately covering disability, especially in relation to menstrual health. Contributing factors include limited data and inadequate training for service providers on disability-inclusive WASH services, resulting in a low awareness of the challenges faced by people with disabilities.[41]

---

[39] M.K. Sharma and R. Adhikari. 2002. Effect of School Water, Sanitation, and Hygiene on Health Status Among Basic Level Students in Nepal. *Environmental Health Insights*. 16.

[40] UNICEF and WHO. 2021. *Progress on Household Drinking Water, Sanitation and Hygiene, 2000–2020: Five Years into the SDGs*.

[41] J. Wilbur et al. 2021. Are Nepal's Water, Sanitation and Hygiene and Menstrual Hygiene Policies and Supporting Documents Inclusive of Disability? A Policy Analysis. *International Journal for Equity in Health*. 20 (157).

Consequently, people with disabilities in Nepal often struggle to access and use water sources and WASH facilities due to physical barriers. These obstacles, present both at home and in public spaces, frequently force individuals to rely on family members or others for assistance.[42] This reliance creates significant challenges for people with disabilities when accessing essential public services.

People with Diverse Sexual Orientations, Gender Identities and Expressions, and Sexual Characteristics

People with diverse sexual orientation, gender identity, gender expression, and sex characteristics (SOGIESC), including transgender and third-gender individuals, encounter specific barriers to accessing public WASH facilities. Although Nepal's first gender-inclusive public toilet opened in 2012, people with diverse SOGIESC still experience discrimination related to their gender identity, sexual orientation, or appearance. This can lead to exclusion; denial of access to WASH facilities; and instances of physical, verbal, or sexual violence—particularly affecting transgender and third-gender individuals.[43]

"[Before the renovations,] we were compelled to go to the nooks and corners for urinating and defecating. Although I could not see anything through my own eyes, I felt like others saw me. Recalling that I feel embarrassed even now."

– Anjali, a 15-year-old visually impaired student.

Source: WaterAid. 2022. Reaching the Unreached: Ensuring Dalit Communities in Nepal Have Decent Toilets. *WaterAid Blog.* 17 November. https://washmatters.wateraid.org/blog/reaching-the-unreached-ensuring-dalit-communities-in-nepal-have-decent-toilets.

"Most public toilets designated for males only have urinals, and they don't have dustbins for the disposal of sanitary pads… On the other hand, I am subjected to hate and disgust in toilets designated for females… I want to be able to enter the 'male' bathroom, but I hope it also has pads, a dustbin, and enough water."

– Zion Magar, a young transgender man in Kathmandu

Source: A. Ghimire. 2022. The Politics of Public Toilets. *The Kathmandu Post.* 14 June. https://kathmandupost.com/national/2022/06/14/the-politics-of-public-toilets.

Policy Context

The Government of Nepal prioritizes the WASH sector, supported by the Ministry of Water Supply and the Department of Water Supply and Sewerage Management. Article 35 (4) of the Constitution of Nepal and Clause 3 of the Water and Sanitation Act 2022 ensure every citizen's right to safe drinking water and sanitation, underscoring the importance of GESI and Nepal's commitment to universal and equitable WASH access.

Nepal's first National Sanitation and Hygiene Master Plan (2011–2017) was released in 2011. This strategic document outlined the country's vision, goals, and objectives for achieving universal access to sanitation and hygiene by 2030. The main objective was to achieve universal sanitation and hygiene access by 2017, including becoming open defecation-free, which the Government of Nepal declared on 30 September 2019.

---

42   SNV Nepal and CBM Australia. 2019. *WASH Experiences of People with Disabilities.*
43   P. Boyce et al. 2018. Transgender-Inclusive Sanitation: Insights from South Asia. *Waterlines.* 37 (2). pp. 102–117.

The master plan has been delayed but is expected to be superseded by the Nepal Water Supply, Sanitation and Hygiene Sector Development Plan (SDP), 2016–2030,[44] led by the Ministry of Water Supply and Sanitation. The SDP, a long-term development plan, acknowledges that although the previous master plan catalyzed a sanitation movement in Nepal (including significant funding allocations), progress had stagnated. The SDP recognizes remaining significant challenges in reducing disparities in access and inequality in services, aiming to ensure WASH services reach all and are effectively utilized, without discriminating between rural and urban areas. The SDP is implemented alongside the Nepal National Water Plan, 2005–2027 and is broken down into three phases: (i) universal access to basic WASH services and improved service levels (2016–2020); (ii) improved service levels (medium or high), functionality, and sustainability improvement (2021–2025); and (iii) improved service levels and impact assessment (2026–2030).

For public buildings, the SDP offers three indicators for institutional sanitation:

(i)   All institutions should have user-friendly, clean, hygienic toilets with handwashing stations and proper waste management facilities.
(ii)   All schools must have child-, gender-, and differently abled-friendly water, toilet, and handwashing (with soap station) facilities, including menstrual hygiene facilities.
(iii)   All institutions should maintain clean and hygienic premises.

Support for the SDP will come from the Water Supply and Sanitation National Policy (still in draft form at the time of writing), which has the vision of "building clean and prosperous society and pollution-free environment through safe, accessible, and easy water supply and sanitation services." This new policy will replace existing rural and urban water and sanitation policies and sit alongside the 2009 National Health Policy to guide all components of WASH in Nepal.

The Water Supply and Sanitation Act 2022, passed in mid-2022, regulates safe drinking water and sanitation services and their establishment, O&M, and management.

Public Water, Sanitation, and Hygiene Facilities

Nepal's National Building Code (NBC) provides clear guidelines for WASH facilities in public buildings. Originally written in 1994 and updated in 2003, the NBC includes a dedicated section (NBC 208) on sanitation and plumbing design requirements. It outlines general guidelines for water supply installation, sewage and wastewater disposal, and rainwater disposal applicable to all residential and public buildings. The NBC aims to ensure an uninterrupted and adequate water supply and safe disposal of sewage, wastewater, and rainwater.

---

[44]   At the time of writing, both the SDP and its supporting policy (the Water Supply and Sanitation National Policy) are in draft form and have not yet been approved by the Government of Nepal.

Specifically, the NBC recommends the following for WASH services and facilities:

(i)   Water storage tanks are required for all buildings without main water connections, providing sufficient pressure and quantity to meet supply needs. Minimum sizes of tanks and minimum standards for access, ventilation, overflow, and other aspects are specified in Part A of NBC 208.

(ii)   Public buildings must provide toilets, ablution taps, handwashing facilities, urinals, and drinking water fountains, with minimum standards for different categories of buildings outlined in Part B of NBC 208.

(iii) Sewage and wastewater should be disposed of into public sewers or, where unavailable, discharged into septic tanks, stabilization ponds, or other approved methods, with minimum standards detailed in Part B of NBC 208.

(iv)   Rainwater from roofs, paved areas, and other open spaces must be collected and disposed of efficiently and quickly, with stormwater disposal systems leading to public drains or natural watercourses, with minimum standards provided in Part C of NBC 208.

Table 5 provides a simplified list of the NBC's recommendations for public WASH facilities.

Table 5: Minimum Water, Sanitation, and Hygiene Recommendations for Public Buildings in Nepal

Female Toilets[a]	Male Toilets[a]	Urinals	Ablution Taps	Handwashing Basins	Drinking Water Fountains
Office Buildings					
1 per 15 persons (and part thereof)	1 per 25 persons (and part thereof)	• 0 for up to 6 persons • 1 for 7–20 persons • 2 for 21–45 persons • 3 for 46–70 persons • 4 for 71–100 persons • For 101–200 persons, add at a rate of 3% • For more than 200 persons, add at a rate of 2.5%	1 in each toilet cubicle	1 per 25 persons	1 per 100 persons (minimum of 1 per floor)

continued on next page

Table 5 *continued*

Female Toilets[a]	Male Toilets[a]	Urinals	Ablution Taps	Handwashing Basins	Drinking Water Fountains
Factories					
• 1 for 1–12 persons • 2 for 13–25 persons • 3 for 26–40 persons • 4 for 41–57 persons • 5 for 58–77 persons • 6 for 78–100 persons • For 101–200 persons, add at a rate of 5% • For more than 200 persons, add at a rate of 4%	• 1 for 1–15 persons • 2 for 21–45 persons • 3 for 36–65 persons • 4 for 66–100 persons • For 101–200 persons, add at a rate of 3% • For more than 200 persons, add at a rate of 2.5%	• 0 for up to 6 persons • 1 for 7–20 persons • 2 for 21–45 persons • 3 for 46–70 persons • 4 for 71–100 persons • From 101–200 persons, add at a rate of 3% • For more than 200 persons, add at a rate of 2.5%	1 in each toilet cubicle	1 per 25 persons	1 for every 100 persons (minimum of 1 per floor)
Art Galleries, Libraries, and Museums					
• 1 per 100–200 persons • More than 200 persons, add at a rate of 1 per 159 persons (and part thereof)	• 1 per 200–400 persons • More than 400 persons, add at a rate of 1 per 250 persons (and part thereof)	1 per 50 persons	1 in each toilet cubicle	• Men: 1 per 200 persons (and part thereof) • Women: 2 per 200 persons (and part thereof)	1 per 100 persons (and part thereof)
Hospitals (Wards)					
1 per 8 beds (and part thereof)			1 in each toilet cubicle	2 per 30 beds, add 1 per additional 30 beds (or part thereof)	
Schools and Educational Institutions (Nonresidential)					
1 per 25 pupils (and part thereof)	1 per 40 pupils (and part thereof)	1 per 20 pupils (and part thereof)	1 in each toilet cubicle	Minimum of 2 (then 1 per 40 girls and 1 per 60 boys)	1 per 50 pupils (and part thereof)

continued on next page

Table 5 *continued*

Female Toilets[a]	Male Toilets[a]	Urinals	Ablution Taps	Handwashing Basins	Drinking Water Fountains
Schools and Educational Institutions (Residential)[b]					
1 per 6 pupils (and part thereof)	1 per 8 pupils (and part thereof)	1 per 25 pupils (and part thereof)	1 in each toilet cubicle	• 1 per 8 boys (and part thereof) • 1 per 6 girls (and part thereof)	1 per 50 pupils (and part thereof)
Warehouses, Fruit and Vegetable Markets					
Minimum of 2 (then 1 per 50 additional persons)		Not less than 2 for every 50 additional persons	Minimum of 2 (then 1 per 50 additional persons)		
Transport Stations and Airports					
Junction stations, intermediate stations, and bus stations					
4 for the first 1,000 persons, then 1 per additional 1,000 persons	3 for the first 1,000 persons, then 1 per additional 1,000 persons	4 for the first 1,000 persons, then 1 per additional 1,000 persons			
Terminal stations and bus stations					
5 for the first 1,000 persons, then 1 per additional 1,000 persons	4 for the first 1,000 persons, then 1 per additional 1,000 persons	6 for the first 1,000 persons, then 1 per additional 1,000 persons			
Domestic airports					
• Minimum of 4 • For 200 persons: 8 • For 400 persons: 15 • For 600 persons: 20 • For 800 persons: 26 • For 1,000 persons: 29	• Minimum of 2 • For 200 persons: 5 • For 400 persons: 9 • For 600 persons: 12 • For 800 persons: 16 • For 1,000 persons: 18	• Minimum of 2 • For 200 persons: 6 • For 400 persons: 12 • For 600 persons: 16 • For 800 persons: 20 • For 1,000 persons: 22			
International airports					
• For 200 persons: 10 • For 600 persons: 20 • For 1,000 persons: 29	• For 200 persons: 6 • For 600 persons: 12 • For 1,000 persons: 18	• For 200 persons: 8 • For 600 persons: 16 • For 1,000 persons: 22			

[a] Some toilets may use Western-style flush toilets with tanks, bowls, and seats, if desired.
[b] Additional bathing facilities is required for 1 for every 6 girls (and part thereof) and 1 for every 8 boys (and part thereof).

Source: Government of Nepal. 2003. *National Building Code.*

Water, Sanitation, and Hygiene in Health Care Facilities

In Nepal, the National Standards for WASH in Health Care Facilities in Nepal (2018) emphasize that WASH is essential for delivering quality health services and people-centered care. These standards ensure facilities are clean and services are safe, thereby boosting trust, demand, and staff morale and performance. Based on WHO's Essential Environmental Health Standards in Health Care (2008), these national standards identify basic and advanced guidelines for water, sanitation, hand hygiene, and waste disposal across various levels of health facilities. They provide excellent guidance on achieving basic and advanced (levels I and II) WASH in health care facilities. Table 6 summarizes the basic standards, while Table 7 covers the advanced standards.

The 2018 national standards also address crucial GESI-responsive components, ensuring that toilets are child- and disability-friendly, menstrual health is supported with appropriate waste bins, and privacy and safety are reinforced through locks and lighting. Additionally, WASH facilities can be further improved by incorporating gender-neutral or unisex toilets and providing in-cubicle water sources (either running water or a dedicated bucket with drainage) for personal cleaning or washing reusable menstrual materials.

Table 6: Basic Water, Sanitation, and Hygiene Standards for Health Care Facilities in Nepal

Water	Sanitation	Hand Hygiene	Waste Disposal
Water from an improved source is available on the premises.	Improved toilets are usable, separated for patients and staff, accommodate menstrual hygiene management, and meet the needs of people with limited mobility.	Hand hygiene materials, either a basin with water and soap or alcohol hand rub, are available at points of care and toilets.	Waste is safely segregated into at least three bins in the consultation area, and sharps and infectious waste are treated and disposed of safely.

Source: Government of Nepal. 2018. *National Standards for WASH in Health Care Facilities (HCF) in Nepal.*

Table 7: Advanced Water, Sanitation, and Hygiene Standards for Health Care Facilities in Nepal

Location	Advanced (Level I)	Advanced (Level II)
Compound (outside main building)	• 1 male block (1 toilet, 3 urinals, double handwashing basin) • 1 female block (3 toilets, double handwashing basin) • 1 unisex accessible toilet	• Toilets in a ratio of 1:100 for men and 2:100 for women • Urinals in a ratio of 1:50 for men • Washbasins in a ratio of 1:100 for men and 2:100 for women
Registration or waiting area	1 common toilet	• 1 male block (1 toilet, 3 urinals, double handwashing basin) • 1 female block (3 toilets, double handwashing basin)

continued on next page

Table 7 *continued*

Location	Advanced (Level I)	Advanced (Level II)
Outpatient department (each department or block)	• 1 male block (2 toilets, 3 urinals, double handwashing basin) • 1 female block (5 toilets, double handwashing basin) • 1 unisex accessible toilet	• Toilets in a ratio of 1:100 for men and 2:100 for women • Urinals in a ratio of 1:50 for men • Washbasins in a ratio of 1:100 for men and 2:100 for women • 1 male staff toilet with handwashing basin • 1 female staff toilet with handwashing basin
Inpatient ward (each ward)	• 1 male block (1 toilet, 3 urinals, 1 shower, double handwashing basin) • 1 female block (4 toilets, 1 shower, double handwashing basin) • 1 unisex accessible toilet	• 1 toilet for every 8 beds for men and 1 for every 6 beds for women • 1 shower for every 8 beds • 2 handwashing basins for up to 30 beds and 1 additional for every 30 beds • 1 urinal for every 12 beds • Bathing facility in toilet for those with special needs
Consultation room	Attached toilet with handwashing basin	Wheelchair friendly and with handles
Laundry room	1 common toilet with handwashing basin	1 toilet with separate washbasins for men and women
Ultrasound sonography room	One toilet with handwashing basin nearby	Wheelchair friendly and with handles
Operation room	Scrub area in the operation room with an attached sluice	
Pathology block	One toilet with handwashing basin nearby	Wheelchair friendly and with handles
Labor or postpartum room	One toilet with handwashing basin	
Delivery room	Sluice with a separate exit route adjoining the delivery room	
Handwashing basin at all service points	1 handwashing basin	
Mortuary	1 common toilet	Bathroom
Laboratory	1 washbasin for each compartment	

Source: Government of Nepal. 2018. *National Standards for WASH in Health Care Facilities (HCF) in Nepal.*

Water, Sanitation, and Hygiene in Schools

In Nepal, the adoption of the school-led total sanitation approach began in 2005. Subsequent policies have included the Guidelines on School Led Total Sanitation (2006, supported by UNICEF), the National School Health and Nutrition Strategy (2006), the National Framework of Child-Friendly Schools (2010), and the School Sector Reform Plan (2009–2015). By 2011, the Government of Nepal committed to ensuring all schools had functioning, child-friendly, and gender-separated toilets with facilities for menstrual hygiene management. These policies culminated in the Wash in Schools Procedure (2018), which adopts the Three Star Approach.

Despite these initiatives, as of 2020, the Ministry of Education, Science and Technology reported that only 372 of 1,554 schools across the country were using the Three Star Approach, with fewer than 20 achieving two- or three-star ratings. This indicates the need for further improvements in GESI-responsive WASH facilities for all students, teachers, and staff.

While the standards outlined in Nepal's NBC broadly apply to all public buildings, schools require additional considerations. The WASH in School Procedure, developed in 2018, provides guidance on the design and O&M of WASH facilities in schools, offers monitoring and evaluation recommendations, and outlines the roles and responsibilities of stakeholders.

The procedure sets the following standards for WASH in schools:

(i)   A sufficient and reliable water supply that is safe for drinking, handwashing, and other hygiene purposes must be provided, with at least 20 liters per day for each person.
(ii) Separate toilets for boys and girls must be available, each with lockable doors, maintaining a ratio of one toilet for every 20 girls and one for every 25 boys, while ensuring that school grounds and the surrounding area are open defecation-free.
(iii) Toilets should accommodate the accessibility needs of people with disabilities.
(iv) Female toilets should provide a safe and private place for changing menstrual materials and include facilities for washing and drying used materials.
(v)   Handwashing facilities should be located near toilets and other areas where students gather, equipped with clean water, soap, and drying methods, and maintaining a ratio of one station for every 50 students.

Importantly, these standards are inherently GESI-responsive as they cater to the diverse accessibility and usage needs of girls and people with disabilities, including provisions for menstruation hygiene management.

The WASH in School Procedure mandates that schools establish a functioning school WASH coordination committee responsible for the planning, implementation, and O&M of WASH facilities and activities. This committee should comprise the following members:

(i)   a school management committee member, nominated by the chair–president,
(ii)   the head teacher or principal,
(iii)   a parent–teacher association representative,
(iv)   a health facility representative from within the school's service area,
(v)   a child club representative,
(vi)   a ward committee representative, and
(vii)   a WASH focal teacher.

The school WASH coordination committee plays an important role in ensuring GESI-responsive WASH facilities at schools by planning and implementing activities, allocating the WASH budget, and monitoring and reviewing the school's WASH facilities. The committee also conducts an annual self-assessment of the school's WASH status to be submitted to the school management committee and the municipality's social development committee.

CASE STUDY—FIJI

Overview

Since the 2000s, Fiji has seen notable improvements in its WASH situation. As of 2022, 95% of the population had access to at least basic drinking water supplies, with 91% coverage in rural areas. However, only 42% of the total population and 27% of the rural population used improved water supplies. Significant improvements in sanitation have been noted. Basic sanitation service access increased from 80% in 2000 to 93% by 2020. By 2023, 49% of the total population had access to safely managed sanitation services, 87% had access to basic hygiene facilities, and less than 1% practiced open defecation (footnote 16).

Health care facilities in Fiji rank among the best in the South Pacific for WASH, though challenges in sanitation persist. By 2021, 69% of health care facilities had at least basic water services (of which 55% was piped and 30% was improved non-piped), but 15% had no water service at all. A 2020 census of WASH in health care facilities in Fiji found that 10% of hospitals and 20% of nonhospital facilities lacked onsite drinking water, requiring patients to bring their own. In 2021, 42% of health care facilities provided basic hygiene services, but 24% had none. Only 9% of health care facilities had at least basic sanitation, while 79% had limited sanitation facilities. Of the health care facilities, 39% offered gender-segregated sanitation facilities, 17% had improved toilets with menstrual hygiene facilities, and 33% had improved toilets accessible to people with limited mobility. In most countries, sanitation services are better in hospitals than in nonhospital facilities, but the opposite has been found in Fiji—none of the country's hospitals met the criteria for basic sanitation services in 2021 (footnote 35).

In Fiji's schools, achievements in WASH vary. As with health care facilities, most schools had at least basic water services (87% in 2021). Only 9% of schools had no water services at all. Meanwhile, 76% of schools had access to at least basic sanitation services, and just 6% had no sanitation services. There is minimal difference in service levels for water and sanitation between primary and secondary schools. However, there is a large gap in hygiene services (a facility with water and soap) between primary (76%) and secondary schools (45%). Overall, 70% of Fiji's schools provide students with basic hygiene services (footnote 36).

How Gender and Social Identity Affect Access to Water, Sanitation, and Hygiene in Fiji

Children

Inadequate WASH facilities at schools can contribute to lower school attendance rates in Fiji, particularly following major disasters such as cyclones. After Tropical Cyclone Winston in 2016, almost 500 schools suffered damage or destruction, including toilet facilities, delaying students' return to school for weeks or even months until repairs were made or temporary structures were built.[45]

Women and Girls

A UNICEF study for the school year 2015–2016 found that managing menstruation is challenging for female students in Fiji. Schoolgirls often leave school early because of pain or the inability to change menstrual materials, exacerbated by the inconsistent and unpredictable availability of supplies (water, soap, toilet paper, and sanitary pads), especially following disasters. Respondents also reported that they struggle to concentrate in class or participate in group activities such as sports when menstruating because they are too worried about staining their school uniforms (footnote 25). Additionally, another study found that menstruating girls face teasing and bullying by boys at school, impacting their participation in class and extracurricular activities, and sometimes leading to absenteeism.

"They will tease us [about menstruation]. And sometimes they will just come to you and will be like, 'I know what you have in your bag.' And through that they take money. You give me this and I will keep it as a secret. For some girls they find it fun, but some they find it embarrassing and some of them don't even come back to school for the next day."

– An out-of-school girl in urban Fiji

Source: Y. Mohamed et al. 2018. A Qualitative Exploration of Menstruation-Related Restrictive Practices in Fiji, Solomon Islands and Papua New Guinea. *PLoS ONE*. 13 (12).

People with Diverse Sexual Orientation, Gender Identity, Gender Expression, and Sex Characteristics

Fiji's Constitution forbids discrimination based on sexual orientation, gender identity, and gender expression. Despite this, after Tropical Cyclone Winston in 2016, people with diverse SOGIESC faced difficulties in securing emergency shelter and longer-term housing and WASH solutions. For example, some transgender women reported discomfort using the same shower and toilet facilities as cisgender women (likely because of fear of discrimination or violence), while several same-gender couples felt discriminated against in housing assistance compared with heterosexual couples. Reports of people with diverse SOGIESC being blamed for "causing" the cyclone were also common, increasing their vulnerability during the initial response phase.[46]

---

45 Save the Children Fiji. WASH. http://www.savethechildren.org.fj/wash/.
46 Edge Effect, Rainbow Pride Foundation Fiji, and Oxfam Australia. 2016. *Down by The River: Addressing the Rights, Needs and Strengths of Fijian Sexual and Gender Minorities in Disaster Risk Reduction and Humanitarian Response.*

Policy Context

The 2013 Fiji Constitution enshrines the rights of all Fijians to water and sanitation, requiring the government to ensure access to clean and safe water in adequate quantities (Article 36) and accessible and adequate sanitation (Article 35). Children's rights to sanitation are also highlighted in Article 41. In addition, the Constitution states that every person has the right to a clean and healthy environment.

The Government of Fiji's national strategy to address urban and rural water challenges is outlined in the long-term National Development Plan (NDP) for 2017–2036 and the 5-year NDP (2017–2021). These plans, marking an important development in Fiji's strategic planning, are the first cross-sector NDPs, incorporating several Sustainable Development Goals (SDGs), including those related to water and sanitation. By 2030, the goal is for 100% of the population to have access to piped and safe drinking water and 70% to centralized sewerage systems. The medium-term Water Supply and Sanitation Sector Development Plan (SDP), 2016–2021, focuses on sustainability and aims to provide safe drinking water and improved sanitation services to every Fijian household, supporting Fiji's climate resilience initiatives.

In 2021, the Rural Water and Sanitation Policy was launched. Initially developed and implemented in 2012, the policy was revised to better reflect constitutional rights to water and sanitation and to meet the targets of the NDP. Covering rural areas and maritime islands outside the metered urban and peri-urban water reticulation systems (managed by the Water Authority of Fiji), this policy aims to provide accessible, safe, affordable, and sustainable drinking water and sanitation services.[47] These services are designed to be

(i)   available 24 hours a day, 7 days a week;
(ii)   affordable and sustainable;
(iii)   compliant with the Fiji National Drinking Water Quality Standards and the Environment Management Act 2005;
(iv)   competently planned, designed, and installed;
(v)   effectively maintained and managed; and
(vi)   user-friendly and environmentally safe.

Additionally, the Rural Water and Sanitation Policy emphasizes the importance of addressing the specific needs of women and girls and those in vulnerable situations, reflecting the language of the SDGs.

Public Water, Sanitation, and Hygiene Facilities

Fiji's National Building Code (NBC), ratified as the Public Health (National Building Code) Regulations 2014, governs WASH provisions in both private and public buildings. It covers sections on sanitary facilities (NF2), water supply plumbing (NF5), and sanitary plumbing and drainage for public buildings and group dwellings (NF6) (Table 8).

---

[47]   While the NDP, 2017–2036 sets a target of 70% access to central sewerage systems by 2031, the 2021 Rural Water and Sanitation Policy specifies a slightly different goal, targeting 60% coverage among the rural population by 2031.

Developed in response to disastrous cyclones in the mid-1980s and largely based on Australian and New Zealand standards, the NBC is currently being updated to incorporate sustainable building designs and enhance climate resilience, in line with the 2021 Climate Change Act.

Table 8: Minimum Water, Sanitation, and Hygiene Requirements for Public Buildings in Fiji

Class of Building	Users	Number of Toilets 1 Up to	2 Up to	Each Extra	Number of Urinals 1 Up to	2 Up to	Each Extra	Number of Washbasins 1 Up to	2 Up to	Each Extra
3, 5, 6, and 9 (e.g., hotels, group dwellings, offices, shops)	Employees:									
	Male	20	40	20	25	50	50	60	120	60
	Female	15	30	15				60	120	60
7 and 8 (e.g., warehouses, laboratories, factories)	Employees:									
	Male	20	40	20	25	50	50	30	60	60
	Female	15	30	15				30	60	60
6 (shops, department stores > 900 square meters)	Patrons:									
	Male	500	2,400	1,200	600	1,200	1,200	1,000	4,000	2,000
	Female	300	600	1,200				1,000	4,000	2,000
6 (e.g., restaurants, cafes, bars) and 9a (outpatient health care facilities)	Patrons:									
	Male	50	200	250	50	200	100	50	200	250
	Female	30	70	80				50	200	250
9a (inpatient health care facilities)	Resident patients:									
	Male		16	8				16	32	16
	Female		16	8				16	32	16
	Additional WASH facilities required: one shower for every 8 (or part thereof) patients.									
9b (schools)	Employees:									
	Male	20	40	20	25	50	50	30	60	30
	Female	15	30	15				30	60	30
	Students:									
	Male	30	70	70	30	70	40	60	140	140
	Female	20	40	30				60	140	140
	Additional WASH facilities required: a minimum of one shower each for male and female students.									
9b (early childhood centers)	Children		30	15					30	15
	Additional WASH facilities required: one shower.									
9b (e.g., sporting venues, theaters, cinemas, art galleries, churches)	Participant:									
	Male	20	40	20	10	20	10	20	40	20
	Female	15	30	15				20	40	20
	Spectators or patrons:									
	Male	250	500	500	100	200	100	250	500	500
	Female	75	250	250				250	500	500
	Additional WASH facilities required: 1 shower for every 10 (or part thereof) participants.									

Source: Government of Fiji. 2004. *Public Health (National Building Code) Regulations 2004.*

Fiji's NBC sets specific standards for accommodating the WASH needs of people with disabilities in class 3 buildings (such as group dwellings and hotels) and class 5, 6, 7, and 9 buildings with floor areas covering more than 1,000 square meters (such as offices, shops, health care facilities, and schools). The ratios are identified per 100 people (Table 9), and the NBC notes that in all cases, female facilities must include "adequate means for the disposal of sanitary towels."

Table 9: Minimum Water, Sanitation, and Hygiene Requirements for People with Disabilities in Fiji

Number of People	Minimum Number of Facilities for People with Disabilities
1–100	• One unisex facility, or • One toilet and washbasin for each sex
101–200	• Two unisex facilities, or • One toilet and washbasin for each sex and one unisex facility
More than	• Two unisex facilities, or one toilet and washbasin for each sex and one unisex facility; and • One additional unisex facility or one toilet and washbasin for each sex for every additional 100 people

Source: Government of Fiji. 2004. *Public Health (National Building Code) Regulations 2004.*

Water, Sanitation, and Hygiene in Health Care Facilities

In addition to the standards set by Fiji's NBC, WASH in health care facilities is guided by the *Guidelines for Climate-Resilient and Environmentally Sustainable Health Care Facilities in Fiji*. Released in 2020 by the Ministry of Health and Medical Services, these guidelines adopt the WHO's Guidance for Climate-Resilient and Environmentally Sustainable Health Care Facilities. As the first of their kind in the Pacific region, they aim to ensure that the country's health care infrastructure is "extra resilient" against frequent cyclones, sea-level rise, and extreme weather events.

Fiji's guidelines are to be used in conjunction with its NBC and other relevant building and development regulations. They outline measures on sustainable building materials and sites, water use and conservation, energy efficiency (including the use of solar photovoltaic systems), chemical management, and waste management. The guidelines also provide climate-resilient and environmentally sustainable health care facility checklists for key groups of interventions, including WASH and health care waste management, as well as assessment tools and mitigation strategies.

Water, Sanitation, and Hygiene in Schools

The Ministry of Education, National Heritage, Culture and Arts set minimum WASH standards for schools in 2012. These standards identify minimum service levels for WASH promotion in schools across Fiji and provide assessment checklists for water quality, water quantity, toilets, cleaning and waste disposal, and hygiene promotion (Table 10). Notably, the standards address the gendered impacts of inadequate WASH facilities in schools and propose sample designs for school toilet cubicles and gender-segregated toilet blocks, including designs for accessible toilets for wheelchair users.

Table 10: Minimum Standards for Water, Sanitation, and Hygiene in Schools in Fiji

Item	Minimum Standard
Water	
Potable water (schools may opt to provide water to children or to ask children to bring potable water from home)	
Full-day pupil	1 liter/day
Boarding pupil	2 liters/day
Teacher/other staff member	4 liters/day
Non-potable water	
Full-day pupil (in areas where water is not readily available or not available)	5 liters/day
Full-day pupil (in areas where water is available)	10 liters/day
Boarding pupil	20 liters/day
Sanitation[a]	
Girls	
Toilets	1 toilet for every 20 girls or part thereof up to 200 girls 1 additional toilet for every 25 girls or part thereof up to 300 girls 1 additional toilet for every 33 girls or part thereof over 300 girls
Handwashing stations	1 handwashing station with tap and soap per 50 girls
Boys	
Toilets	1 toilet for every 33 boys or part thereof up to 200 boys 1 additional toilet for every 50 boys or part thereof over 200 boys
Urinals	1 urinal per 50 boys
Handwashing stations	1 handwashing station with tap and soap per 50 boys
Staff	
Toilets	Separate toilets for each sex 1 toilet for every 20 staff or part thereof of either sex Minimum of 2 cubicles for women and 1 for men
Handwashing stations	1 handwashing station with tap and soap per 20 staff
Children who use wheelchairs	
Toilets	For newly constructed toilet facilities, 1 toilet for children who use wheelchairs

[a] Special schools for children with disabilities should have one toilet cubicle for every 15 children, including one toilet for wheelchair users for each toilet block and per sex.

Source: Government of Fiji. 2012. *Minimum Standards on Water, Sanitation and Hygiene (WASH) in Schools Infrastructure.*

The standards specify that both girls and boys must have equal access to adequate sanitation facilities in schools, ensuring privacy for all. Facilities for girls and boys must be separate, each equipped with handwashing stations and adequate visual, noise, and odor separation. Additionally, staff toilets must have separate facilities for men and women, with adequate privacy.

Menstrual hygiene management, also mentioned in the standards, is addressed by requiring at least one shower room for both male and female students and placing one sanitary bin in each girl's toilet cubicle and an additional sanitary bin per block of toilets. Menstrual products such as pads should be readily available in schools to support students' needs, and incinerators for the disposal of used menstrual materials may be installed, pending approval from local authorities.

6 CONCLUSION

Achieving universal access to GESI-responsive public WASH facilities in developing Asia and the Pacific, including schools and health care facilities, demands substantial commitment from government agencies. This includes sustained funding and human resources to enhance and maintain these WASH facilities, ensuring they remain safe, hygienic, and well-managed. Effective O&M plans, backed by adequate funding and community engagement, are crucial for long-term sustainability.

While progress has been made, significant challenges persist, particularly in sanitation. Government policies, standards, and guidelines provide a solid framework for action and emphasize GESI principles, offering clear implementation steps. Strengthening GESI efforts further, integrating the human right to water and sanitation with a human rights-based approach (HRBA), ensures the protection of rights and dignity for all, particularly the most marginalized and vulnerable people.

In essence, the objective of SDG 6 for universal safe water and sanitation embodies a commitment to inclusivity, proactively addressing the specific needs of women, girls, other genders, people with disabilities, older people, and other marginalized and vulnerable groups. Addressing these needs from the outset, rather than as an afterthought, can significantly improve the effectiveness of public WASH facilities.

APPENDIXES

Appendix 1: Checklist—Enabling Environment Assessment

This checklist can be used to assess the status of water, sanitation, and hygiene (WASH) sector policies, budgets, and plans at the national and/or subnational levels.

Checklist 4: National and Subnational Enabling Environment Assessment

Component		Yes/No
Policy, Legislation, and Guidelines		
Is there a WASH sector policy in place?		
Does the policy incorporate gender equality principles?		
Does the policy incorporate social inclusion principles?		
Is the policy being implemented?		
Is there WASH legislation in place?		
Does the legislation incorporate gender equality principles?		
Does the legislation incorporate social inclusion principles?		
Is the legislation being implemented?		
Are there guidelines or standards in place for		
	WASH?	
	WASH in schools?	
	WASH in health care facilities?	
If no overarching WASH guidelines exist, are there stand-alone guidelines or standards for the provision of the following services and facilities:		
	Water?	
	Sanitation?	
	Hygiene?	
	Menstrual health and hygiene?	

continued on next page

Checklist 4 *continued*

Component	Yes/No
Planning and Budgeting	
Are planning and budgeting mechanisms used for the WASH sector?	
Do national WASH sector plans and budgets take subnational plans and budgets into consideration and vice versa?	
Are community members actively engaged in WASH sector planning and budgeting?	
Are national and subnational WASH data used in planning and budgeting?	
Is there an up-to-date (i.e., covering the current year) multiyear WASH sector plan (strategy) in place?	
If yes: Is there a costed budget for the plan?	
Does the plan take into account the needs for capital investment as well as ongoing O&M?	
Does the plan incorporate principles of gender and social inclusion?	
Is the plan being implemented?	
Is implementation on track?	
Is budget expenditure on track?	
Data, Monitoring, and Evaluation	
Does the WASH sector plan include a monitoring framework with targets, indicators, and methods of measurement?	
Is the plan being monitored and evaluated?	
Are the results of monitoring and evaluation of the WASH sector plan presented to the national Parliament?	
Are up-to-date WASH data available at the national level?	
Are up-to-date WASH data available at the subnational level?	
Are data on WASH in schools available?	
Are data on WASH in health care facilities available?	
Are WASH data made available to global monitoring mechanisms, such as the SDGs and the WHO/UNICEF JMP?	
Is WASH data made publicly available?	

JMP = Joint Monitoring Programme for Water Supply, Sanitation and Hygiene; O&M = operation and maintenance; SDG = Sustainable Development Goal; UNICEF = United Nations Children's Fund; WASH = water, sanitation, and hygiene; WHO = World Health Organization.

Appendix 2: Additional Resources

Asian Development Bank (ADB). 2023. *Addressing Menstrual Health in Urban, Water, and Sanitation Interventions in the Pacific: Practitioner Guide.* https://www.adb.org/sites/default/files/publication/891411/addressing-menstrual-health-pacific-practitioner-guide.pdf.

United Nations Entity for Gender Equality and the Empowerment of Women (UN Women). 2022. *Intersectionality Resource Guide and Toolkit: An Intersectional Approach to Leave No One Behind.* https://www.unwomen.org/sites/default/files/2022-01/Intersectionality-resource-guide-and-toolkit-en.pdf.

United Nations Children's Fund (UNICEF). 2019. *Global Framework for Urban Water, Sanitation and Hygiene.* https://www.unicef.org/media/63941/file/Global%20Framework%20for%20Urban%20Water,%20Sanitation%20and%20Hygiene.pdf.

UNICEF. 2019. *Guidance on Menstrual Health and Hygiene.* https://www.unicef.org/media/91341/file/UNICEF-Guidance-menstrual-health-hygiene-2019.pdf.

UNICEF. 2021. *Water, Sanitation and Hygiene (WASH): A Guidance Note for Leaving No One Behind (LNOB).* https://www.unicef.org/media/102136/file/LNOB-in-WASH-Guidance-Note.pdf.

UNICEF and Deutsche Gesellschaft für Internationale Zusammenarbeit. 2013. *Field Guide: The Three Star Approach for WASH in Schools.* https://globalhandwashing.org/wp-content/uploads/2015/03/UNICEF_Field_Guide-3_Star-Guide1.pdf.

UNICEF and World Health Organization (WHO). 2022. Hand Hygiene Acceleration Framework Tool. https://www.who.int/publications/m/item/the-hand-hygiene-acceleration-framework-tool-(hhaft).

WaterAid and Government of Australia, Department of Foreign Affairs and Trade. 2018. Accessibility and Safety Audits. 14 August. https://washmatters.wateraid.org/publications/accessibility-and-safety-audits.

WaterAid; Government of the United Kingdom, Department for International Development; and Sanitation and Hygiene Applied Research for Equity. 2015. *Menstrual Hygiene Matters: Training Guide for Practitioners.* https://washmatters.wateraid.org/sites/g/files/jkxoof256/files/MHM%20training%20guide_0.pdf.

WaterAid, Water and Sanitation for the Urban Poor, and UNICEF. 2018. *Female-Friendly Public and Community Toilets: A Guide for Planners and Decision Makers.* https://washmatters.wateraid.org/sites/g/files/jkxoof256/files/female-friendly-public-and-community-toilets-a-guide.pdf.

Water for Women Fund; Edge Effect; and Government of Australia, Department of Foreign Affairs and Trade. 2020. *Sexual and Gender Minorities and COVID-19: Guidance for WASH Delivery.* Guidance Note: COVID-19. https://www.waterforwomenfund.org/en/learning-and-resources/resources/KL/WfW_EdgeEffect_Guidance-Note_COVID-19-WASH-SGM-Inclusion-FINAL.pdf.

Water for Women Fund and Sanitation Learning Hub. 2021. *Gender Equality and Social Inclusion Self-Assessment Tool: Facilitation Guide for WASH Project Managers, Researchers and Self-Assessment Facilitators—Working Towards Transformation in Inclusive Water, Sanitation and Hygiene.* https://opendocs.ids.ac.uk/opendocs/bitstream/handle/20.500.12413/16810/Gender%20Equality%20 and%20Social%20Inclusion%20Self-Assessment%20Tool%20-%20Water%20for%20Women%20 Fund%20and%20SLH.pdf.

World Health Organization (WHO). 2022. *Sanitary Inspections for Sanitation.* https://cdn.who.int/ media/docs/default-source/wash-documents/sanitation/sanitation-inspection-forms/sanitary-surveys-for-sanitation-complete-2022-11-15-reduced.pdf.

WHO. 2019. *Water, Sanitation, and Hygiene in Health Care Facilities: Practical Steps to Achieve Universal Access to Quality Care.* https://iris.who.int/bitstream/handle/10665/311618/9789241515511-eng.pdf.

WHO and UNICEF. 2022. *Water and Sanitation for Health Facility Improvement Tool (WASH FIT): A Practical Guide for Improving Quality of Care Through Water, Sanitation and Hygiene in Health Care Facilities (Second Edition).* https://iris.who.int/bitstream/handle/10665/353411/9789240043237-eng. pdf.

GLOSSARY

Gender	The socially constructed characteristics of women, men, and others, such as norms, roles, and relationships between them. Gender varies from society to society and can change over time. The concept of gender includes five important elements: relational, hierarchical, historical, contextual, and institutional. While most people are born either male or female, they are taught appropriate norms and behaviors, including how they should interact with others of the same or opposite sex within households, communities, and workplaces. When individuals or groups do not match established gender norms, they often face stigma, discriminatory practices, or social exclusion.
Gender equality and social inclusion (GESI)	A concept that addresses unequal power relations experienced by people on the grounds of gender and other social statuses, such as wealth or assets, class or caste, job or source of income, ethnicity or indigeneity, disability, citizenship, location, language, sexuality, religion or beliefs, health status, and agency.
GESI-responsive	Approaches that recognize, consider, and respond to issues of GESI in design, implementation, and monitoring, as well as address the barriers to the participation and representation of marginalized and vulnerable groups.
Gender-neutral	Not specific to or associated with a particular gender. For example, a gender-neutral toilet can be accessed and used by all genders, while a gender-neutral policy is not aimed at any particular gender and is assumed to the affect all genders equally. Gender-neutral facilities are sometimes referred to as "unisex" facilities.
Intersectionality	A way of understanding how gender, race, ethnicity, sexual orientation, gender identity, disability, class, and other identities "intersect" to create different experiences and forms of discrimination. The creator of the term, Kimberlé Crenshaw, summarizes intersectionality as "a way of thinking about identity and its relationship to power."
Intersex	People who are born with genetic, hormonal, or physical sex characteristics that do not conform to medical norms for "male" or "female" bodies.
Marginalized groups	Groups discriminated against or excluded in terms of participating in mainstream social, cultural, and economic activities. As a result, marginalized groups may be confined to the edges of society and have limited or no access to basic services and opportunities.

Menstrual materials or products	Any materials or products used to absorb, collect, and/or dispose of menstrual blood.
Menstrual health	A state of complete physical, mental, and social well-being and not merely the absence of disease or infirmity, in relation to the menstrual cycle.
Menstruation	Bleeding from the uterus that happens about once a month as a normal part of the menstrual cycle. Another word for menstruation is "period."
People with disabilities	All people who have long-term physical, mental, intellectual, and/or sensory impairments, which, in interaction with various barriers, may hinder their full and effective participation in society on an equal basis with others.
People who menstruate	All people who experience the biological process of menstruation. Women and girls make up the majority of people who menstruate. However, other people may also menstruate, including transgender men, nonbinary individuals, third-gender individuals, and intersex individuals.
Public water, sanitation, and hygiene (WASH) facilities	WASH facilities such as toilets, handwashing stations, and water access points, located in public buildings, public spaces, and other places frequently accessed by the public.
Sex	The biological and physiological characteristics that define humans as female or male, such as reproductive organs, chromosomes, and hormones. These sets of biological characteristics are not mutually exclusive, as there are individuals who possess both, but these characteristics tend to differentiate humans as females or males.
Sexual orientation, gender identity, gender expression, and sex characteristics (SOGIESC)	An abbreviation used to describe sexual orientation, gender identity and expression, and sex characteristics collectively for the purposes of law, policy, and programming, with individual terms as follows: **Sexual orientation:** A person's romantic or sexual attraction to another person, including heterosexual, homosexual or gay, lesbian, bisexual, pansexual, asexual, and same-sex attracted, among others. **Gender identity:** A person's deeply held internal and individual feeling of gender. **Gender expression:** The way in which a person externally expresses their gender or how they are perceived by others. **Sex characteristics:** A person's primary and secondary sex characteristics, including an individual's sex chromosomes, hormones, reproductive organs, genitals, and breast and hair development.

Third gender	An umbrella term for people who consider themselves as neither male nor female. Some countries officially recognize third-gender people and offer third gender as a gender identity on important documents such as passports.
Transgender	An umbrella term for people whose gender identity is different from that which was legally assigned to them at birth. Transgender is an adjective, so a person who is transgender can be referred to as a transgender person.
Universal design	Principles that accommodate the needs of all people regardless of age, ability, and other social identities. In the context of public WASH facilities, universal design means that people of all identities and abilities can access and use WASH facilities safely, hygienically, and with dignity.

www.ingramcontent.com/pod-product-compliance
Lightning Source LLC
LaVergne TN
LVHW071452180726
843512LV00018B/1362